Genes and Antigens in Cancer Cells –
The Monoclonal Antibody Approach

Contributions to Oncology
Beiträge zur Onkologie

Vol. 19

Series Editors
S. Eckhardt, Budapest; *J. H. Holzner,* Wien;
G. A. Nagel, Göttingen

S. Karger · Basel · München · Paris · London · New York · Tokyo · Sydney

Proceedings of the 4th International Expert Meeting of the
Deutsche Stiftung für Krebsforschung, Bonn, June 27–29, 1983

Genes and Antigens in Cancer Cells –
The Monoclonal Antibody Approach

Volume Editors
G. Riethmüller, München; *H. Koprowski,* Philadelphia;
S. von Kleist, Freiburg i. Br.; *K. Munk,* Heidelberg

59 figures (13 fig. in colour) and 42 tables, 1984

S. Karger · Basel · München · Paris · London · New York · Tokyo · Sydney

Contributions to Oncology
Beiträge zur Onkologie

National Library of Medicine, Cataloging in Publication
    Genes and antigenes in cancer cells – the monoclonal antibody: symposium,
    Heidelberg, 27–29 June, 1983 / volume editors, G. Riethmüller . . . [et al.].
    – – Basel; New York: Karger, 1984.
    (Contributions to oncology = Beiträge zur Onkologie; v. 19)
    Includes index.
    1. Antibodies, Monoclonal – congresses   2. Neoplasms – immunology – congresses
    I. Riethmüller, G. (Gert)   II.
    W1 BE444N v. 19 [QZ 200 G327 1983]
    ISBN 3-8055-3843-X

Drug Dosage
    The authors and the publisher have exerted every effort to ensure that drug selection and dos-
    age set forth in this text are in accord with current recommendations and practice at the time
    of publication. However, in view of ongoing research, changes in government regulations,
    and the constant flow of information relating to drug therapy and drug reactions, the reader
    is urged to check the package insert for each drug for any change in indications and dosage
    and for added warnings and precautions. This is particularly important when the recom-
    mended agent is a new and/or infrequently employed drug.

All rights reserved.
    No part of this publication may be translated into other languages, reproduced or utilized in
    any form or by any means, electronic or mechanical, including photocopying, recording,
    microcopying, or by any information storage and retrieval system, without permission in
    writing from the publisher.

©   Copyright 1984 by S. Karger AG, P.O. Box, CH-4009 Basel (Switzerland)
    Printed in Germany by Ernst Kieser GmbH, D-8900 Augsburg
    ISBN 3-8055-3843-X

Contents

Preface VII

Meuer, S. C.; Cooper, D. A.; Schlossman, S. F.; Reinherz, L.
(Boston, Ma.): Clonal Analysis of Human Immunoregulatory
and Effector T Lymphocytes in Viral Infection 1

Croce, C. M. (Philadelphia, Pa.): Chromosome Translocations and
Oncogene Activation in Burkitt Lymphoma 21

Mölling, K.; Donner, P.; Bunte, T.; Greiser-Wilke, I. (Berlin):
Molecular Mechanisms of Malignant Transformation
by Oncornaviruses 35

Ginsburg, V.; Fredman, P.; Magnani, J. L. (Bethesda, Md.):
Cancer-Associated Carbohydrate Antigens Detected by
Monoclonal Antibodies 44

Feizi, T. (Harrow): Monoclonal Antibodies Reveal Saccharide
Structures of Glycoproteins and Glycolipids as Differentiation
and Tumour-Associated Antigens 51

Zimmermann, W.; Thompson, J.; Grunert, F.; Luckenbach, G.-A.;
Friedrich, R.; von Kleist, S. (Freiburg, i. Br.): Identification
of Messenger RNA Coding for Carcinoembryonic Antigen 64

Knapp, W.; Majdic, O.; Bettelheim, P.; Liszka, K.; Stockinger, H.
(Vienna): Antigenic Heterogeneity in Acute Leukemia 75

Stein, H.; Gerdes, J.; Lemke, H.; Burrichter, H.; Diehl, V.; Gatter, K.;
Mason, D. Y. (Oxford): Hodgkin's Disease and So-Called
Malignant Histiocytosis: Neoplasms of a New Cell Type 88

Bander, N. H. (New York, N. Y.): Renal Cancer – A Model System
for the Study of Human Neoplasia 105

Hellström, K. E.; Hellström, I.; Brown, J. P.; Larson, S. M.;
Nepom, G. T.; Carrasquillo, J. A. (Washington, D. C.):
Three Human Melanoma-Associated Antigens and
Their Possible Clinical Application 121

Johnson, J. P.; Holzmann, B.; Kaudewitz, P.; Riethmüller, G. (Munich):
Monoclonal Antibodies Directed Against Transformation
Related Antigens in Melanoma 132

Brüggen, J.; Bröcker, E.-B.; Suter, L.; Redmann, K.; Sorg, C.
(Münster): Comparative Analysis of Melanoma-Associated
Antigens in Primary and Metastatic Tumour Tissue 139

Osborn, M.; Debus, E.; Altmannsberger, M.; Weber, K. (Göttingen):
Uses of Conventional and Monoclonal Antibodies
to Intermediate Filament Proteins in the Diagnosis
of Human Tumours 148

Herlyn, M.; Blaszcyk, M.; Koprowski, H. (Philadelphia, Pa.):
Immunodiagnosis of Human Solid Tumours 160

Mach, J.-P.; Buchegger, F.; Haskell, C.; Carrel, S. (Épalinges/
Lausanne); *Forni, M.; Ritschard, J.; Donath, A.* (Geneva):
In Vivo Localization of Polyclonal and Monoclonal Anti-CEA
Antibodies in Human Colon Carcinomas 171

Sears, H. F.; Herlyn, D.; Koprowski, H.; Wen Shen, J.
(Philadelphia, Pa.): Immunotherapy of Gastrointestinal
Malignancies with a Murine IgG 2A Antibody 180

Preface

Oncology in its clinical and theoretical ramifications has undergone an intimate merger with immunology in recent years. Though for quite some time, tumour immunology has existed as a discipline more by wishful thinking than by generation of hard data, the recent development of several new techniques has revolutionized the approach to malignant transformation. Modern immunology has contributed not only exciting new tools for the precise analysis and localization of molecules in whole cell, but it has also provided new concepts of gene organization and gene expression, vital to our understanding of cell differentiation. As demonstrated in the analysis of the immunoglobulin genes and their interaction with oncogenes, unexpected links have emerged between possible control mechanisms of malignant growth and genuine immunological differentiation processes.

It was on the broad field of cell differentiation that the "Deutsche Krebshilfe" decided to convocate a number of prominent scientists in order to exchange the latest information and newest ideas. Although there was a strong motivation within the "Deutsche Krebshilfe" to concentrate on applied research, the organizers were completely free to compose the final programme.

As it turned out the range of topics treated in the conference showed a considerable breadth reaching from molecular genetics and theoretical immunology to clinical applications such as radioimaging or therapy of human tumours with monoclonal antibodies.

With the title, "Genes and Antigens in Cancer Cells", it became clear that not all important points of current research could be covered. However, during the conference presentations, many starting points for wider discussion were given. The conference began with addresses on central issues of current immunology. *Meurer's* contribution provided the evidence for clontypical expression of surface glycoproteins on cloned T lymphocytes. It was clear the combination of two approaches – the monoclonal antibody technique and the in vitro growth of cloned T cells – now presents the unresolved question of antigen recognition by T cells amenable to analysis.

The frequently invoked heterogeneity of tumour cells was elegantly shown by *Knapp* in an analysis on human leukemia cells with monoclonal antibodies. As demonstrated by *Stein,* the classification of Hodgkin's disease and large cell lymphoma is about to appear under a new perspective based on the presence of a hitherto unknown antigen defined by monoclonal antibody Ki-1. One of the central issues of the conference, the role of oncogenes, was discussed by *Croce,* who presented a review on his results on the effects of oncogenes when translocated to the heavy chain locus in Burkitt lymphoma cells. The possible role of human T leukemia virus in the causation of human leukemia and acquired immunodeficiency was discussed by *Gallo*. As described by *Lane* and *Mölling* monoclonal antibodies are particularly useful in defining proteins encoded by viral genes in several tumour virus systems.

The intricacies of carbohydrate chemistry on tumour cells were dealt with in presentations of *Ginsburg* and *Feizi* who showed how monoclonal antibodies can be applied for determining sequences of sugars in glycolipids and glycoproteins. One of the major glycoproteins on tumour cells is the carcinoembryonal antigen, the isolation of the messenger RNA of CEA was presented by *Zimmermann*.

A whole section of the conference was devoted to the cartography of surface antigens on human melanoma cells. *Hellström, Brüggen* and *Johnson* gave their view on the current state of art in this exciting field of human tumour immunology.

It is reasonable to expect that in the foreseeable future we will have a good inventory of surface molecules on this particular human tumour. A similar study on kidney carcinoma cells was presented by *Bander* from the Sloan Kettering Institute, who reported on a large library of monoclonal antibodies defining several differentiation antigens in kidney cancer cells and in normal kidney cells.

The last half day of the meeting was devoted to diagnostic and therapeutic applications of monoclonal antibodies.

Preface IX

New tumour markers in serum, originally defined by monoclonal antibodies on tumour cell surfaces, were presented by *Herlyn*. The recent classification of cytoskeleton proteins according to major cell differentiation lineages has opened up a new avenue to tumour typing. As demonstrated by *Osborn* monoclonal antibodies are particularly useful for typing intermediate filament proteins in tumour cells of uncertain histogenetic origin.

Vitetta gave a review on her work on therapy of mouse lymphoma tumour cells with monoclonal antibodies. *Mach* summarized his data on imaging using a variety of monoclonal antibodies. As to therapy, *Sears* presented the first data on monoclonal antibodies in colon carcinoma patients. With the outlook on the existing experimental models and the first studies in patients, it became quite clear that with the advent of monoclonal antibodies the era of a new pharmacology is about to unfold!

The organizers are greatly indebted to the "Deutsche Krebshilfe" for financial support and to several people who played a key role behind the scenes. Our thanks are particulary due to the staff of the "Deutsche Krebshilfe" whose devotion to the task and effectiveness in coping with numerous, varied arrangements did so much in assuring the final success of the conference.

Munich, December 1983 *G. Riethmüller*

Clonal Analysis of Human Immunoregulatory and Effector T Lymphocytes in Viral Infection*

S. C. Meuer, D. A. Cooper, S. F. Schlossman, L. Reinherz

Division of Tumor Immunology, Dana-Farber Cancer Institute, Boston, Mass., USA

Introduction

With the development of monoclonal antibodies to human T lymphocyte surface determinants, it has become clear that two functionally distinct human T cell subsets exist, and that these subpopulations express unique cell surface molecules. Thus, the T4+ subset provides inducer/helper function for T-T, T-B and T-macrophage interactions whereas the T8+ subset principally functions in a suppressive mode. Moreover, although both subsets of cells proliferate to alloantigen in mixed lymphocyte culture, the vast majority of cytotoxic effector function is usually detected in the T8+ population. It is important to note that the development of cytotoxicity by T8+ cells in general requires interactions with T4+ cells or their soluble products. In contrast, only a minor component of cytotoxic effector function resides within the T4+ subset and this is maximal when T4+ cells alone are sensitized in MLC [1–3].

In the face of antigenic challenge by viral, fungal, protozoan or bacterial pathogens, these cells are activated to fulfill their individual functional programs [4–11]. The complexity of cellular immune responses to viral infections has been studied both in vitro and in vivo. For example, in acute mononucleosis, both cytotoxic as well as suppressor T lymphocytes directed against virus-infected B cells are activated, presumably in order to terminate viral replication and B cell hyperreactivity [12–17]. At the same time, a specific antibody response directed at viral gene products is induced by T lymphocytes.

* This work was supported by NIH Grants IRO1 AI 19807-01 and RO1 NS 17182 and DFG Grant Me 693/1-1.

Together, these mechanisms are believed to be sufficiently protective to result in cessation of clinical disease activity.

The in vitro generation of clonal T cell populations which are specific for EBV infected autologous B lymphocytes now affords the opportunity to directly study the individual T cell elements involved in the above immune responses [14, 17, 18]. Thus, in the present report we characterized a series of human T cell clones specific for autologous EBV transformed B lymphoblastoid cells with regard to immunoregulatory and effector function. In the results we demonstrate that three discrete T cell populations involved in the regulation of B cell immunoglobulin production can be defined at the clonal level. In addition, cytotoxic effector lymphocytes derived from both major human T cell subsets, T4+ and T8+, are generated in this in vitro system which exhibit dual specificity for virally encoded or induced antigens in the context of different classes of MHC gene products.

Results and Discussion

Surface Phenotypes and Function of Regulatory T Cell Clones

A series of semi-solid agar derived human T cell colonies [19, 20] generated against an autologous EBV transformed B cell line (termed Laz 509) was analyzed for their capacity to regulate Ig production by autologous B lymphocytes. To this end, 1.5×10^5 autologous peripheral blood mononuclear cells (PBMC) and 0.5×10^5 cells derived from individual T cell colonies were incubated with PWM for 7 days at 37 °C. Subsequently, culture supernatants were harvested and quantitated in a solid phase radioimmunoassay (RIA). Of 22 T cell colonies tested, 3 induced a significant increase in B cell IgG secretion, 11 strongly inhibited IgG secretion and 8 colonies had no effect. It was found that all inducer T cell clones reacted with anti-T4 whereas clones which inhibited Ig production from PBMC were either anti-T4 or anti-T8 reactive (table I). A number of the above colonies were cloned by limiting dilution technique, expanded in liquid culture and characterized with a series of monoclonal antibodies directed at T cell surface structures. Three representative immunoregulatory T cell clones termed $AT4_{II}$, $AT4_{IV}$ and $AT8_{III}$ were chosen for in depth analysis.

Table I. Distribution of regulatory T cell colonies following stimulation with autologous EBV transformed B lymphocytes

	T8+	T4+
Enhancement	0	3
Inhibition	8	3

Figure 1 indicates the surface phenotypes of $AT4_{II}$, $AT4_{IV}$ and $AT8_{III}$ as determined by indirect immunofluorescence with monoclonal antibodies and goat anti-mouse FITC on an Epics V cell sorter. As shown, all clones express the mature T cell surface glycoproteins T1, T3 and T12. In addition, each population is Ia+, a characteristic of activated T lymphocytes. Whereas $AT4_{II}$ and $AT4_{IV}$ are reactive with anti-T4 and unreactive with anti-T8, $AT8_{III}$ is reactive with anti-T8 and unreactive with anti-T4. In no case did these or other clones coexpress anti-T4 and anti-T8 reactivity.

The regulatory influence of $AT4_{II}$, $AT4_{IV}$ and $AT8_{III}$ on IgG synthesis was examined in 4 separate experiments. The latter were performed over a period of several months with consistent results indicative of the clones' functional stability. Given the observed variation in the absolute amount of IgG synthesized among individual experiments, the results are expressed as percentage of control for simplicity of comparison (100% = autologous PBMC + PWM). As shown in

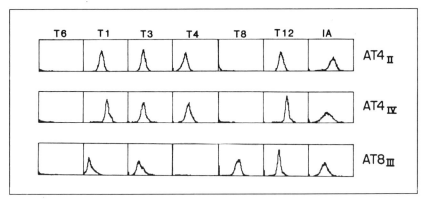

Fig. 1. Cell surface expression of differentiation antigens by clones $AT4_{II}$, $AT4_{IV}$ and $AT8_{III}$. T cell clones were incubated with saturating concentrations of one or another monoclonal antibody in ascites form and goat anti-mouse F(ab')$_2$ FITC. Analysis was performed on 10000 cells/sample employing an Epics V cell sorter.

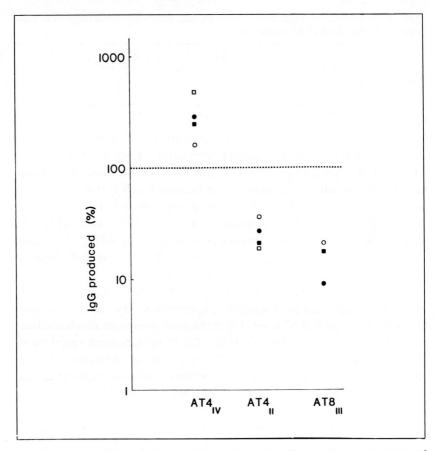

Fig. 2. Regulatory activities of clones $AT4_{II}$, $AT4_{IV}$ and $AT8_{III}$. 1.5×10^5 autologous PBMC and 0.5×10^5 cloned cells were incubated for 7 days with PWM (1:500). Supernatants were then analyzed for total IgG in a solid phase RIA. Results are given as percentage of the control: ---, 100% (1.5×10^5 PBMC stimulated with PWM, 1:500). The various symbols (■, ●, □, ○) represent the results of four individual experiments performed over a period of several months.

figure 2, clone $AT4_{IV}$ significantly enhanced the secretion of IgG production by PBMC in each experiment performed (160–430% of control). In contrast, both $AT4_{II}$ and $AT8_{III}$ exhibited strong suppressive activity (17–35% and 9–22%, respectively). The inductive activity of $AT4_{IV}$ was due to a direct effect on B lymphocytes since this clone by itself triggered isolated B cells to produce Ig (data not shown).

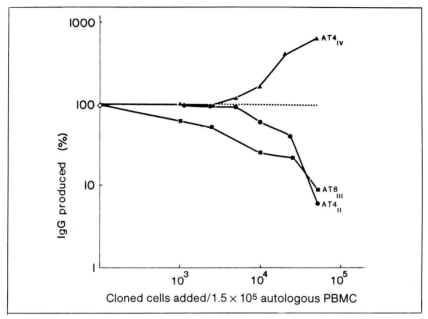

Fig. 3. Effects of varying numbers of cloned cells on IgG production. 1.5×10^5 autologous PBMC and varying numbers of $AT4_{II}$, $AT4_{IV}$ and $AT8_{III}$ cells were incubated with PWM (1:500) for 7 days. Supernatants were analyzed for IgG in a solid phase RIA. ●, $AT4_{II}$; ▲, $AT4_{IV}$; ■, $AT8_{III}$. Note that the results are expressed on a log vs. log plot.

In the above experiments, a fixed number of cloned cells had been added to each culture. We next investigated the effects on immunoregulatory function by varying the number of $AT4_{II}$, $AT4_{IV}$, and $AT8_{III}$ cells when added to a constant number of autologous PBMC (1.5×10^5/well) in the presence of PWM. As shown in figure 3, $5-10 \times 10^3$ cells from clone $AT4_{IV}$ were sufficient to provide help for IgG synthesis. Moreover, IgG production increased in a dose dependent fashion with increasing numbers of $AT4_{IV}$ cells. In a reciprocal way, the addition of increasing numbers of either $AT4_{II}$ or $AT8_{III}$ cells depressed IgG synthesis by PWM stimulated PBMC. Greater than or equal to 10^3 $AT8_{III}$ cells or $>5 \times 10^3$ $AT4_{II}$ cells resulted in a diminution of Ig production. These dose dependent results excluded the possibility that the above differential immunoregulatory effects of these clones were due to trivial variations in culture conditions which they might have produced at a single concentration.

Dissection of Clonal Suppressor Activities

In earlier studies employing heterogeneous peripheral blood T cell populations, it was demonstrated that generation of suppression required an interaction between a radiosensitive T4+ inducer cell and a radiosensitive T8+ suppressor cell. The former was necessary to activate the latter to become a T8+ suppressor effector [21–24]. It was, therefore, of importance to determine whether $AT4_{II}$ and $AT8_{III}$ clones themselves had a direct suppressive effect on IgG synthesis or alternatively, whether they mediated suppression via additional effector T cell populations. To this end, fresh autologous T lymphocytes and fractionated T4+ and T8+ T cells were added in varying combinations to B lymphocytes with one or another T cell clone. The mix-

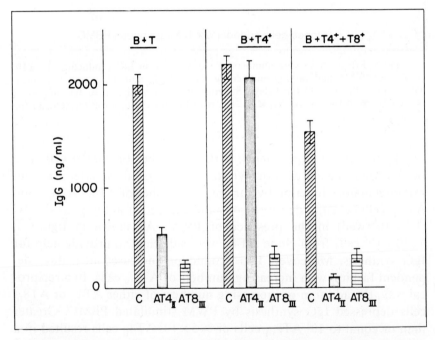

Fig. 4. Dissection of suppressor inducer and suppressor effector function. 0.5×10^5 B lymphocytes were incubated in various combinations with E+ lymphocytes (1×10^5), T4+ lymphocytes (0.5×10^5), T8+ lymphocytes (0.3×10^5) and clones $AT4_{II}$ or $AT8_{III}$ (0.5×10^5) for 7 days in the presence of PWM (1:500). Supernatants were analyzed for total IgG in a solid phase RIA.

ture was then stimulated with PWM for 7 days and IgG synthesis assayed as above. As shown in figure 4, the mixture of PWM stimulated B cells and unfractionated T cells resulted in 2000 ng/ml of IgG during the 7-day culture. Addition of either $AT4_{II}$ or $AT8_{III}$ to this mixture resulted in marked suppression of IgG production (530 and 230 ng/ml, respectively). In contrast, very different results were obtained when $AT4_{II}$ and $AT8_{III}$ cells were individually added to a combination of purified autologous T4+ T cells and B cells. Whereas $AT4_{II}$ induced little or no diminution in Ig production, $AT8_{III}$ reduced IgG production by 75%.

The above finding suggested that the suppressive effect of the $AT4_{II}$ clone was mediated via the T8+ cells present in the unfractionated T cell population. To test this possibility, $AT4_{II}$ cells were added to a mixture of autologous B cells, T4+ T cells and T8+ T cells stimulated by PWM. As shown in figure 4, addition of fresh T8+ T cells to the peripheral mononuclear mixture reconstituted the suppressive effect of the $AT4_{II}$ clone (B+T4+T8 = 96 ng/ml vs. B+T4 = 2000 ng/ml). Although not shown, irradiation of the unfractionated T cell population with 1500 rad also markedly diminished the suppressive effect of the $AT4_{II}$ cells. From these results, it would appear that the $AT8_{III}$ clone represents a suppressor effector cell whereas the $AT4_{II}$ clone, in contrast, induces a T8+ pre-suppressor cell to become a suppressor effector (i. e., inducer of suppression). In addition to emphasizing the distinct subpopulations involved in the generation of suppression, the above experiment excluded the possibility that suppression mediated by $AT4_{II}$ resulted from nonspecific cytotoxicity. If the latter were the case, then $AT4_{II}$ would have killed the autologous T4+ plus B cell mixture, resulting in reduced rather than the observed normal IgG secretion.

Regulatory Effects of Supernatants

In order to determine whether soluble factors derived from $AT4_{II}$, $AT4_{IV}$ or $AT8_{III}$ could mediate some of the clone's regulatory function, individual T cell clones were incubated at 1.5×10^6 cells/ml with an equal number of irradiated stimulator cells (Laz 509, 5000 rad) for varying periods of time. Subsequently, supernatants were harvested, added to fresh autologous PBMC in the presence or ab-

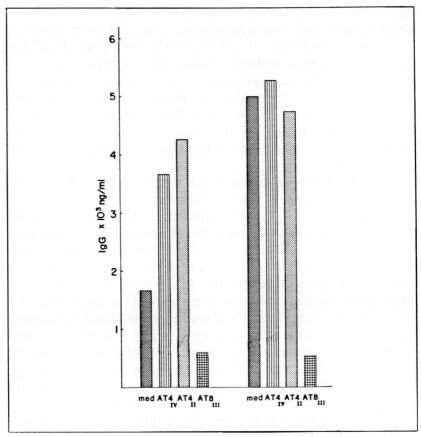

Fig. 5. Regulatory activities of supernatants derived from stimulated clones $AT4_{II}$, $AT4_{IV}$ and $AT8_{III}$. 1.5×10^5 autologous PBMC were incubated with one or another supernatant (final concentration: 20%) for 7 days without PWM (left group of columns) or with PWM (1:500) (right group of columns).
▨, medium control; ▥, $AT4_{IV}$; ▨, $AT4_{II}$; ▦, $AT8_{III}$.
Supernatants were analyzed for total IgG in a solid phase RIA.

sence of PWM during a 7-day culture period and then IgG secretion quantitated. Figure 5 demonstrates the results of one of a series of experiments of this type. As shown, PBMC spontaneously produce approximately 2000 ng/ml of IgG during the 7-day culture and this was enhanced to 5000 ng/ml upon PWM stimulation. Supernatants from both $AT4_{IV}$ and $AT4_{II}$ provided helper function for Ig production by the unstimulated PBMC. In contrast, the $AT8_{III}$ supernatant

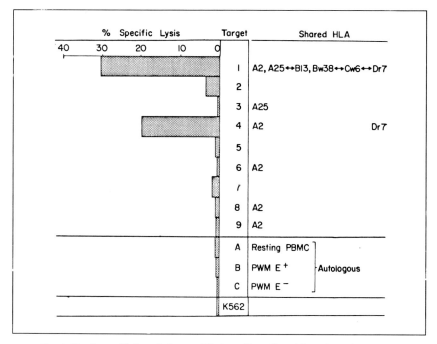

Fig. 6. Dual specificity of clone $AT4_I$ in cell mediated lympholysis. $AT4_I$ was incubated at a 15:1 E/T ratio with a panel of ^{51}Cr labelled EBV transformed HLA typed lymphoblastoid B cell lines [1–9], autologous resting and PWM activated cells and the human standard NK target, K562. The results represent mean values of triplicate cultures in a standard CML assay. Target [1] is the autologous B cell line to which $AT4_I$ had been generated.

inhibited Ig production in the same unstimulated PBMC population. In addition, $AT8_{III}$ markedly suppressed Ig production from the PWM stimulated PBMC. In contrast, supernatants from the two AT4 clones did not detectably enhance or diminish PWM stimulated Ig synthesis.

The inductive and suppressive effects of the $AT4_{IV}$ and $AT8_{III}$ supernatants corresponded to the activities that the respective clones themselves mediated upon addition to PBMC. In both cases, the activities were detectable in supernatants following 48–72 h of stimulation but could not be detected in supernatants harvested after 96 h. Similarly, the fact that the supernatants from the antigen stimulated $AT4_{II}$ clone contained a helper factor which enhanced IgG synthesis by resting PBMC was consistent with the effects mediated by the clone itself in the absence of PWM (data not shown). In contrast, the inability of the $AT4_{II}$ supernatant, unlike the $AT4_{II}$ clone, to suppress Ig synthesis by the PWM stimulated PBMC, suggests that the suppressor inducer factor, if it existed, was not stable under these experimental circumstances.

Clonal Reactivity with JRA Sera and Anti-TQ1

Given the above findings and earlier data indicating that the T4 population could be separated into T4+JRA−TQ1− inducer and T4+JRA+TQ1+ suppression inducer subpopulations, the reactivity of the T cell clones with JRA autoantibodies (present in sera of some patients with juvenile rheumatoid arthritis) and the monoclonal antibody anti-TQ1 was determined [23, 24]. For this purpose, JRA reactivity was assessed in a complement dependent lysis assay whereas anti-TQ1 reactivity was determined by means of indirect immunofluorescence. As shown in table II, both $AT4_{II}$ and $AT8_{III}$ were reactive with JRA antisera whereas $AT4_{IV}$ was unreactive. This was of interest in light of the fact that the former two clones were involved in mediation of suppression whereas the latter was not. Moreover, table II indicates that unlike the resting PBMC population, all clones including the $AT4_{II}$ T4+ inducer of suppression were unreactive with monoclonal antibody TQ1. The finding that anti-TQ1 did not react with any of the clones tested, including $AT4_{II}$, was somewhat unexpected. One explanation for the latter finding could be that the surface expression of the TQ1 molecule was lost following T cell activation. In this regard, we have recently observed that activation of T4+TQ1+ cells in the autologous MLR results in diminished anti-TQ1 activity when analyzed sequentially.

Cytotoxic Effector Function of EBV Specific T Cell Clones

In the next series of experiments, we characterized the cytotoxic effector function of T cell clones generated in this in vitro system. Thus, a number of soft agar derived colonies were investigated for their capacity to lyse the autologous EBV transformed stimulator B cell line, Laz 509. As outlined in table III, 46 cultures expressed the

Table II. Characteristics of human virus specific T cell clones

Designation	$AT4_{IV}$	$AT4_{II}$	$AT8_{III}$
Regulatory function	Helper	Inducer of suppression	Suppressor effector
Subset derivation	T4+/T8−	T4+/T8−	T4−/T8+
JRA reactivity	−	+	+
Anti-TQ1 reactivity	−	−	−
Effector function (CML)	+	+	−
Genetic restriction of CML	Class II MHC	Class II MHC	N. A.

T8 antigen and 13 cultures expressed the T4 antigen. 43 of 46 T8+ cultures and 12 of 13 T4+ cultures exhibited significant levels of cytotoxicity (≥20% lysis) against the stimulating autologous B cell line, Laz 509. Representative T4+ and T8+ cultures were subsequently cloned by limiting dilution and propagated in continuous culture. Four cytotoxic T4+ clones ($AT4_I$–$AT4_{IV}$) and two cytotoxic T8+ clones ($AT8_I$, $AT8_{II}$) will be the subject of the following experiments. Representative phenotypes of such clones are given in figure 1 ($AT8_I$ and $AT8_{II}$ had an identical phenotype as $AT8_{III}$).

Dual Specificity of EBV Specific T Cell Clones

To determine the specificity of the T4+ and T8+ CTL clones, we tested their ability to lyse other EBV transformed lymphoid B cell target cells [1–9] of known HLA-A, B, C and DR phenotypes. The results of such a characterization for the T4+ clone, $AT4_I$, is given in figure 6. As shown, $AT4_I$ is cytotoxic for the autologous B lymphoblastoid line, Laz 509 [1] and one other allogeneic B lymphoblastoid line, designated panel target [4], which shares the HLA-DR7 antigen with Laz 509 [1] but, in contrast, is not cytotoxic for the 7 EBV lym-

Table III. Distribution of cytotoxic T cell colonies following stimulation with autologous EBV transformed B lymphocytes

	T3+	T4+	T8+
Total	59	13	46
Cytotoxic	55	12	43
Noncytotoxic	4	1	3

phoblastoid lines lacking DR7. This result suggests that the $AT4_I$ clone recognizes an epitope of the DR7 molecule on target cells. Moreover, given the fact that $AT4_I$ is not cytotoxic for autologous DR7 resting lymphocytes or syngeneic pokeweed mitogen activated T cell or B cell blasts and that the latter are known to express Ia antigens (fig. 6, A–C), it would appear that this clone views DR7 as well as an additional target antigen, presumably induced and/or encoded by virus.

To further investigate this point, a series of cold target inhibition experiments was performed employing PWM activated autologous lymphocytes. As shown in figure 7, the addition of unlabelled Laz 509 inhibited the ability of clone $AT4_I$ to lyse ^{51}Cr labelled Laz 509 in a dose dependent fashion. Total inhibition of lysis was observed at a ratio of unlabelled/labelled Laz 509 of 20:1. In contrast, autologous PWM activated E+ or E– cells were unable to block the ability of $AT4_I$ to kill ^{51}Cr labelled Laz 509 at any ratio tested. Thus, it would

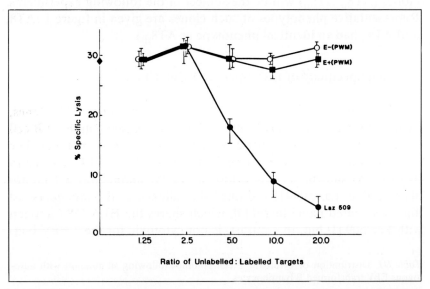

Fig. 7. Cold target inhibition of CML by $AT4_I$ directed at Laz 509. Increasing numbers of unlabelled Laz 509, PWM activated autologous E+ or E– lymphocytes were added to a constant number of ^{51}Cr labelled Laz 509 at the initiation of the CML assay.
◆, lysis in the absence of cold targets (E/T ratio: 15/1); ●, unlabelled Laz 509; ○, unlabelled PWM activated E+; ■, unlabelled PWM activated E–.

appear that $AT4_I$ does not detect an activation antigen on B lymphocytes as induced by PWM.

Although not shown in this article, studies with other clones indicated that they too manifest an analogous "dual specificity", i. e. killed targets expressing an EBV related antigen in the context of HLA or Ia-related molecules. Moreover, each of these CTLs prolifer-

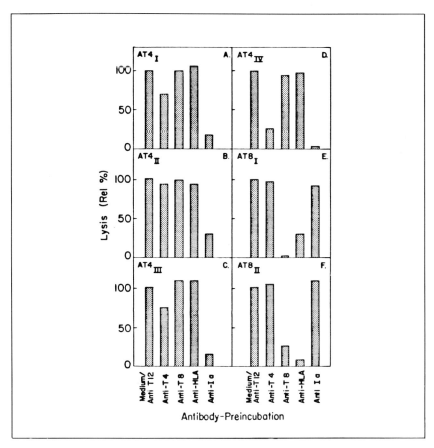

Fig. 8. Inhibitory effects of various antisera on CML by clones $AT4_I$–$AT4_{IV}$ *(A–D)*, $AT8_I$ and $AT8_{II}$ *(E, F)*. Anti-T cell antibodies were incubated with the effector cell populations for 30 min at RT prior to addition of ^{51}Cr labelled autologous EBV transformed targets (final dilutions – anti-T12, 1:300; anti-T4$_B$, 1:300, anti-T8$_A$, 1:500). Antisera to human HLA and Ia antigens were incubated with the targets for 30 min at RT prior to addition of effector cells [final dilutions – w6/32 (anti-HLA), 1:150; p29, 34 (anti-Ia), 1:150]. The E/T ratio in all experiments was 15:1 (comp. fig. 6). Results are representative of four separate experiments.

ated to target cells bearing the appropriate specificities as well. This analysis suggested an identical or similar specificity (HLA-DR7) of three of the T4+ clones (AT4$_I$, AT4$_{II}$, AT4$_{IV}$) and an HLA-A25 specificity of one T8+ clone (AT8$_I$). It should be noted that the stimulatory cell line was homozygous for DR7. In contrast, the specificity of AT4$_{III}$ and AT8$_{II}$ could not be determined with this panel. In addition, none of the clones mediates cytotoxicity via natural killer activity since they are unable to lyse the conventional human NK target, K562.

Differential Inhibitory Effects of Anti-T Cell and Anti-MHC Antibodies

Since previous studies showed that alloreactive cytotoxic T4+ or T8+ clones were directed at different classes of MHC gene products [20, 26], the above findings in the autologous system were not unexpected. To further investigate these differences, we examined the ability of monoclonal antibodies and heteroantisera directed at class I or class II MHC gene products to block cytotoxic effector function by individual T cell clones on the target cell level. As shown in figure 8, antibodies to nonpolymorphic Ia or HLA epitopes had very different effects on the cytotoxicity of T cell clones depending on their T4+ or T8+ phenotype. For example, preincubation of the autologous target, Laz 509, with an anti-p29,34 anti-Ia antiserum resulted in inhibition of killing by all T4 clones AT4$_I$–AT4$_{IV}$ (panels A–D, respectively). In contrast, the same anti-Ia heteroantiserum could not inhibit cytotoxicity by two representative T8+ clones, AT8$_I$ and AT8$_{II}$ (panels E and F).

In a reciprocal fashion, monoclonal antibody w6/32, which is directed at a nonpolymorphic determinant on the alpha chain of HLA-A, B and C antigens strongly inhibited CML by the T8+ clones but not by the T4+ clones. Although not shown, heteroantisera and monoclonal antibodies against β2 microglobulin did not influence the level of cytotoxicity by either T4+ or T8+ clones.

Given the correlation between surface expression of the T4 and T8 surface glycoproteins on the T cell clones themselves and the specificity for class II and class I MHC gene products, it was important to determine whether anti-T4 and anti-T8 antibodies influenced killing function by the respective T effector cell populations. To examine this point, the same cytotoxic T cell clones were preincubated with one or another T cell specific monoclonal antibody (anti-T4, anti-T8, anti-T12) or media prior to the standard CML assay. As shown in figure 8 (panels E and F), anti-T8 antibodies markedly diminished the cytotoxic effector function of both T8+ clones but did not diminish the level of killing by the T4+ clones, $AT4_I-AT4_{IV}$. In contrast, monoclonal anti-T4 antibody markedly inhibited cytotoxicity of the T4 clone $AT4_{IV}$ (panel D), partially inhibited cytotoxicity of the T4 clones $AT4_I$ (panel A) and $AT4_{III}$ (panel C) and had minimal or no effect on T4+ clone $AT4_{II}$ (panel B). These same antibodies did not influence killing by the two T8 clones, $AT8_I$ and $AT8_{II}$. That the inhibitory effects were not simply a function of antibody binding to clonal effector cells was evident from the finding that a monoclonal antibody, anti-T12, directed at a 120KD glycoprotein present on both T4+ and T8+ clones, had no effect on cytotoxicity. In addition, monoclonal antibodies directed at the 20KD T3 molecule present on all clones, blocked cytotoxic effector function as well as antigen induced clonal proliferation (data not shown) which is in line with earlier results regarding the critical importance of the T3 surface structure for antigen specific human T cell function [27-31].

Prior studies in both human and murine systems have indicated that cytotoxic T cell populations specific for virally infected autologous cells demonstrated dual specificity for autologous class I alloantigens and virally encoded determinants. At the clonal level, their effector function could be blocked by antibodies to the T lymphocyte surface antigen T8 and Lyt2 in man and mouse, respectively [14, 32, 33]. The present findings are consistent with the above in that the vast majority of clones are T8+ and presumably recognize EBV related antigens in the context of autologous class I MHC gene products. Moreover, these results are compatible with findings in the peripheral blood of patients with acute infectious mononucleosis, an EBV infection, where there is a marked increase in the T8+ population [12]. Perhaps more importantly, the present study shows that human cytotoxic T lymphocytes directed at autologous EBV transformed target cells can be derived from the T4+ T cell subset as well. In contrast to the „classical" T8+ CTL, these appear to exhibit specificity for a virally induced or encoded antigen in the context of self class II MHC gene products. It es yet to be determined whether the same or different viral determinants associate with class I and class II MHC gene products.

Simultaneous Expression of Regulatory and Cytotoxic Effector Function

As indicated in table II, clones $AT4_{II}$ and $AT4_{IV}$, but not $AT8_{III}$, in addition to their regulatory influence on B cell Ig production, also mediated cytotoxic effector function directed at autologous EBV transformed B lymphoblasts (see below). Thus, it appears to be possible for a single T cell to exhibit effector and regulatory functions simultaneously.

These results are in keeping with recent studies in the murine system suggesting that such clones exist [25]. The presence of an inducer and cytotoxic cell which facilitates an antibody response and simultaneously directs cytotoxic activity at EBV transformed cells may be of considerable protective value to the host. The frequency of such multifunctional clones is not known at present. Given the inability of $AT8_{III}$ to effect cytotoxic function, it is, however, obvious that some clones exclusively mediate regulatory activities in the absence of other effector functions.

Conclusions and Speculations

The precise significance of the T4+ class II specific CTL is not known. Although it represents a minority of CTL effectors generated from the unfractionated T cell population, prior studies indicated that T4+ cells sensitized to antigen in the absence of T8+ cells generated an increased amount of cytotoxicity [2]. This observation raised the possibility that the T8+ population, known to contain suppressor cells, regulated the expression of cytotoxicity within the reciprocal T4+ subset [1, 34]. Thus, while T4+ CTL may not represent the major cytotoxic effector population in the normal situation, they may assume great significance in patients with autoimmune disorders who lack a functional T8+ subset. Moreover, some of these cells could potentially give rise to autoimmune phenomena as a consequence of cytotoxicity directed against exogenous or endogenous antigen in the context of autologous Ia molecules. In this regard, it is known that many patients with systemic lupus erythematosus and multiple sclerosis lack T8+ cells at the time of acute disease exacerbation [35]. In the case of lupus dermatitis, for example, there is selective loss of the Ia+ Langerhans cells from the skin and concomitant T4+ T cell infiltration into the dermis [36]. Likewise, T4+ cells are prominently represented in the acute plaque lesions within brain parenchyma of multiple sclerosis patients where oligodendrocytes are being destroyed [37, 38]. A further understanding of the specificities of T4+ effector function and their potential regulatory effects may help in the elucidation of these and other autoimmune disorders.

Since the above phenotypically and functionally distinct clones were generated to autologous B lymphocytes transformed by EB virus, it is apparent that multiple regulatory and effector populations are activated during the process of viral infection. The ability to generate such homogeneous populations of cells in vitro should provide an important means to study their surface structures and the regulatory molecules which they produce and, therefore, allow detailed insight in immunoregulatory mechanisms which maintain homeostasis. In addition, such studies provide means to further dissect the functional human T cell compartments.

Acknowledgments: The authors wish to thank Ms. *K. A. Fitzgerald*, Mr. *J. C. Hodgdon* and Ms. *R. E. Hussey* for excellent technical assistance and Mr. *J. F. Daley* and *M. H. Levine* for performing Epics V cell sorter analysis.

Summary

Human virus specific T cell clones and monoclonal antibodies directed at their surface structures are used to investigate immunoregulatory and effector functions of T lymphocytes triggered as a consequence of viral infection. Thus following stimulation with autologous EBV transformed B cells at least three phenotypically and functionally distinct T cell populations exhibiting regulatory activity in T-B interactions are defined at the clonal level: Helper, inducer of suppression, and suppressor effector cells. In addition, cytotoxic T cell clones are generated in this system that display dual specificity for MHC gene products as well as antigens induced or encoded by virus. Blocking studies employing monoclonal anti-T cell antibodies as well as anti-MHC antisera support the notion that the T cell subset restricted glycoproteins, T4 and T8, serve as associative recognition elements for, respectively, class II and class I MHC gene products on target cells.

References

1 Reinherz, E. L.; Schlossman, S. F.: The differentiation and function of human T lymphocytes. Cell *19:* 821–827 (1980).
2 Reinherz, E. L.; Kung, P. C.; Goldstein, G.; Schlossman, S. F.: Separation of functional subsets of human T cells by a monoclonal antibody. Proc. natn. Acad. Sci. USA *76:* 4061–4065 (1979).
3 Reinherz, E. L.; Kung, P. C.; Goldstein, G.; Levey, R. H.; Schloßman, S. F.: Discrete stages of human intrathymic differentiation: Analysis of normal thymocytes and leukemic lymphoblasts of T cell lineage. Proc. natn. Acad. Sci. USA *77:* 1588–1592 (1980).
4 Zinkernagel, R. M.; Doherty, P. C.: H-2 compatibility requirement for T cell mediated lysis of targets infected with lymphocytic choriomeningitis virus. Different cytotoxic T cell specificities are associated with structures coded in H-2K or H-2D. J. exp. Med. *141:* 1427 (1975).
5 Biddison, W. E.; Payne, S. M.; Shearer, G. M.; Shaw, S.: Human cytotoxic T cell responses to trinitrophenyl hapten and influenza virus. Diversity of restriction antigens and specificity of HLA linked genetic regulation. J. exp. Med. *152:* 204 (1980).
6 Carney, W. P.; Rubin, R. H.; Hoffman, R. A.; Hansen, W. P.; Healey, K.; Hirsch, M. S.: Analysis of T lymphocyte subsets in cytomegalovirus mononucleosis. J. Immunol. *126:* 2114 (1981).
7 Lamb, J. R.; Eckels, D. C.; Phelan, M.; Lake, P.; Woody, J. N.: Antigen specific human T lymphocyte clones: viral antigen specificity of influenza virus immune clones. J. Immunol. *128:* 1428 (1982).
8 Stobo, J. D.; Paul, S.; Van Scoy, R. E.; Hermans, P. E.: Suppressor thymus-derived lymphocytes in fungal infections. J. clin. Invest. *57:* 319 (1976).
9 Piessens, W. F.; Partono, F.; Hoffman, S. L.; Ratiwayanto, S.; Piessens, P. W.; Palmieri, J. R.; Koiman, I.; Dennis, D. T.; Carney, W. P.: Antigen specific suppressor T lymphocytes in human lymphatic filariasis. N. Engl. J. Med. *307:* 144 (1982).
10 Mehra, V.; Mason, L. H.; Rothman, W.; Reinherz, E. L.; Schlossman, S. F.; Bloom, B. R.: Delineation of a human T cell subset responsible for lepromin induced suppression in leprosy patients. J. Immunol. *125:* 1183 (1980).
11 Van Voorhis, W. C.; Kaplan, G.; Nunes Sarno, E.; Horwitz, M. A.; Steinman, R. M.; Levis, W. R.; Nogueira, N.; Hair, L. S.; Rocha Gattass, C.; Arrick, B. A.; Cohn, Z. A.: The cutaneous infiltrates of leprosy: Cellular characteristics and the predominant T cell phenotypes. New Engl. J. Med. *307:* 1593 (1982).
12 Reinherz, E. L.; O'Brien, C.; Rosenthal, P.; Schlossman, S. F.: The cellular basis for viral induced immunodeficiency: Analysis by monoclonal antibodies. J. Immunol. *125:* 1269 (1980).
13 Tsukuda, K.; Berczi, I.; Klein, G.: Human B lymphocytes activated by Epstein Barr virus (EBV) or by mitogens suppress mitogen induced immunoglobulin production. J. Immunol. *126:* 1810 (1981).
14 Wallace, L. E.; Rickinson, A. B.; Rowe, M.; Epstein, M. A.: Epstein Barr virus specific cytotoxic T cell clones restricted through a single HLA antigen. Nature *297:* 413 (1982).
15 Tosato, G.; Magrath, I.; Koski, I.; Dooley, N.; Blaese, M.: Activation of suppressor T cells during Epstein-Barr virus induced infectious mononucleosis. N. Engl. J. Med. *301:* 1133 (1979).
16 Tosato, G.; Magrath, I. T.; Blaese, R. M.: T cell mediated immunoregulation of

Epstein Barr virus (EBV) induced B lymphocyte activation in EBV-seropositive and EBV-seronegative individuals. J. Immunol. *128:* 575 (1982).

17 Meuer, S. C.; Hodgdon, J. C.; Cooper, D. A.; Hussey, R. E.; Fitzgerald, K. A.; Schlossman, S. F.; Reinherz, E. L.: Human cytotoxic T cell clones directed at autologous virus-transformed targets: further evidence for linkage of genetic restriction to T4 and T8 surface glycoproteins. J. Immunol. *131:* 186–190 (1983).

18 Meuer, S. C.; Cooper, D. A.; Hodgdon, J. C.; Hussey, R. E.; Morimoto, C.; Schlossman, S. F.; Reinherz, E. L.: Immunoregulatory human T lymphocytes triggered as a consequence of viral infection: clonal analysis of helper, suppressor inducer and suppressor effector cell populations. J. Immunol. *131:* 1172 (1983).

19 Sredni, B.; Volkman, D.; Schwartz, R. H.; Fauci, A. S.: Antigen specific human T cell clones: development of clones requiring HLA-DR compatible presenting cells for stimulation in presence of antigen. Proc. natn. Acad. Sci. USA *78:* 1858–1862 (1981).

20 Meuer, S. C.; Schlossman, S. F.; Reinherz, E. L.: Clonal analysis of human cytotoxic T lymphocytes: T4+ and T8+ effector T cells recognize products of different major histocompatibility complex regions. Proc. natn. Acad. Sci. USA *79:* 4590 (1982).

21 Morimoto, C.; Reinherz, E. L.; Borel, Y.; Mantzouranis, E.; Steinberg, A. D.; Schlossman, S. F.: Autoantibody to an immunoregulatory inducer population in patients with juvenile rheumatoid arthritis. J. clin. Invest. *67:* 753 (1981).

22 Morimoto, C.; Distaso, J. A.; Borel, Y.; Schlossman, S. F.; Reinherz, E. L.: Communicative interactions between subpopulations of human T lymphocytes required for generation of suppressor effector function in a primary antibody response. J. Immunol. *128:* 1645 (1982).

23 Reinherz, E. L.; Morimoto, C.; Fitzgerald, K. A.; Hussey, R. E.; Daley, J. F.; Schlossman, S. F.: Heterogeneity of human T4+ inducer T cells defined by a monoclonal antibody that delineates two functional subpopulations. J. Immunol. *128:* 463 (1982).

24 Morimoto, C.; Reinherz, E. L.; Borel, Y.; Schlossman, S. F.: Direct demonstration of the human suppressor inducer subset by anti-T cell antibodies. J. Immunol. *130:* 157 (1983).

25 Widmer, M. B.; Bach, F. H.: Cytolytic T lymphocyte clones with helper cell characteristics (Abstract). Immunobiology *163:* 153 (1982).

26 Krensky, A. M.; Clayberger, C.; Reiss, C. S.; Strominger, J. L.; Burakoff, S. J.: Specificity of OKT4+ cytotoxic T lymphocyte clones. J. Immunol. *129:* 2001–2003 (1982).

27 Reinherz, E. L.; Hussey, R. E.; Schlossman, S. F.: A monoclonal antibody blocking human T cell function. Eur. J. Immunol. *10:* 758–762 (1980).

28 Reinherz, E. L.; Meuer, S. C.; Fitzgerald, K. A.; Hussey, R. E.; Levine, H.; Schlossman, S. F.: Antigen recognition by human T lymphocytes is linked to surface expression of the T3 molecular complex. Cell *30:* 735–743 (1982).

29 Meuer, S. C.; Hussey, R. E.; Hodgdon, J. C.; Hercend, T.; Schlossman, S. F.; Reinherz, E. L.: Surface structures involved in target recognition by human cytotoxic T lymphocytes. Science *218:* 471–473 (1982).

30 Meuer, S. C.; Fitzgerald, K. A.; Hussey, R. E.; Hodgdon, J. C.; Schlossman, S. F.; Reinherz, E. L.: Clonotypic structures involved in antigen specific human T cell function: relationship to the T3 molecular complex. J. exp. Med. *157:* 705–719 (1983).

31 Meuer, S. C.; Acuto, O.; Hussey, R. E.; Hodgdon, J. C.; Fitzgerald, K. A.; Schlossman, S. F.; Reinherz, E. L.: Evidence for the T3-associated 90KD heterodimer as the T cell antigen receptor. Nature *303:* 808–810 (1983).

32 Doherty, P. C.: Surveillance of self: CMI to virally modified cell surface defined operationally by the MHC; in M. Fougereau Dausset (eds.), Progress in Immunology, pp. 563–576 (Academic Press, New York, 1980).
33 Sarmiento, M.; Glasebrook, A. L.; Fitch, F. W.: IgG or IgM monoclonal antibodies reactive with different determinants on the molecular complex bearing Lyt2 antigen block T cell mediated cytolysis in the absence of complement. J. Immunol. 125: 2665–2669 (1980).
34 Reinherz, E. L.; Morimoto, C.; Penta, A. C.; Schlossman, S. F.: Regulation of B cell immunoglobulin secretion by functional subsets of T lymphocytes in man. Eur. J. Immunol. 10: 570–572 (1980).
35 Reinherz, E. L.; Weiner, H. L.; Hauser, S. L.; Cohen, J. A., Distaso, J. A.; Schlossman, S. F.: Loss of suppressor T cells in active multiple sclerosis: Analysis with monoclonal antibodies. New Engl. J. Med. 303: 125–129 (1980).
36 Sontheimer, R. D.; Bergstresser, P. R.: Epidermal Langerhans cells involved in cutaneous lupus erythematosus. J. invest. Derm. 79: 237–243 (1982).
37 Traugott, U.; Reinherz, E. L.; Raine, C. S.: Multiple sclerosis: Distribution of T cell subsets within active chronic lesions. Science 219: 308–310 (1983).
38 Hauser, S. L.; Bhan, A. K.; Gilles, F.; Reinherz, E. L.; Hoban, C. J.; Weiner, H. L.: Monoclonal antibodies against human Ia and monocyte antigens identify subpopulations of human glial cells. J. Neuroimmunol. (in press, 1983).

Dr. med. Stefan C. Meuer, I. Medizinische Klinik und Poliklinik der Johannes Gutenberg-Universität, Langenbeckstraße 1, D-6500 Mainz (FRG)

Chromosome Translocations and Oncogene Activation in Burkitt Lymphoma

C. M. Croce

The Wistar Institute of Anatomy and Biology, Philadelphia, Pa., USA

Introduction

Burkitt lymphoma is a malignancy of B cells which predominantly affects children [1]. Specific chromosome translocations have been observed in this disease [2, 3]. In about 90% of cases a reciprocal translocation between chromosome 8 and 14 is observed [2, 3]; in the remaining 10% the reciprocal translocations involve either chromosomes 2 and 8 or 8 and 22, the breakpoint on chromosome 8 being consistently on band q24 [4, 5]. Since we and others have assigned the genes for immunoglobulin heavy chains and κ and λ light chains to chromosomes 14 [6], 2 [7, 8], and 22 [9] respectively, we have speculated that the human immunoglobulin genes might be involved in the translocations observed in Burkitt lymphomas [9].

In order to determine whether this is the case we produced somatic cell hybrids between mouse myeloma and human Burkitt lymphoma cells with the t(8;14) translocation and examined the hybrid cells for the presence of the chromosomes involved in this translocation and for the presence, rearrangements, and expression of the human immunoglobulin chain genes [10]. The results of those studies indicate that in Burkitt lymphomas with the t(8;14) translocation the chromosome break occurs within the heavy chain locus and that the genes for the variable regions of heavy chain translocate from chromosome 14 to the involved chromosome 8, and that the expressed heavy chain allele is located on the normal chromosome 14 [10].

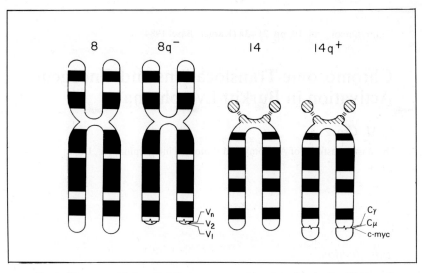

Fig. 1. Diagrams of the t(8;14) chromosome translocation in Burkitt lymphoma cells. The Cμ and Cγ genes are proximal to the breakpoint on chromosome 14, while the V_H gene translocates to the 8q$^-$ chromosome. The human c-*myc* gene on the broken chromosome 8 translocates to the heavy chain locus. Whereas in the Burkitt lymphoma cell lines, CA46 and JD38-IV, the c-*myc* gene is in the same 22-kb BamH1 fragment with the Cμ gene, the c-*myc* gene is not joined to the Cμ gene in P3RH-1 and Daudi Burkitt lymphoma cells.

We have also found that the human homologue, c-*myc*, of the avian myelocytomatosis virus oncogene, v-*myc*, is located on chromosome 8 and it translocates to chromosome 14 in Burkitt lymphomas with the t(8;14) translocation [11]. Further investigations have indicated that in some of the Burkitt lymphoma cells the c-*myc* gene rearranges head-to-head with the Cμ gene while in others the c-*myc* is translocated but is not rearranged within a large BamH1 restriction fragment [11-13]. Figure 1 summarizes the previous findings concerning Burkitt lymphoma with the t(8;14) chromosome translocation.

In the present study we have investigated the expression of the c-*myc* gene in malignant and non-malignant B cells and the mechanisms of oncogene activation in Burkitt lymphomas carrying the variant chromosome translocations.

Results

Transcription of the c-*myc* Oncogene in Burkitt Lymphoma Cells

We have examined the levels of c-*myc* transcripts in Burkitt lymphoma cells, in human lymphoblastoid cells and in HL60 human promyelocytic cells which contain an amplified c-*myc* oncogene which is expressed at a high level [14] by S_1 protection method [15]. As shown in figure 2, Burkitt lymphoma cell lines express high levels of

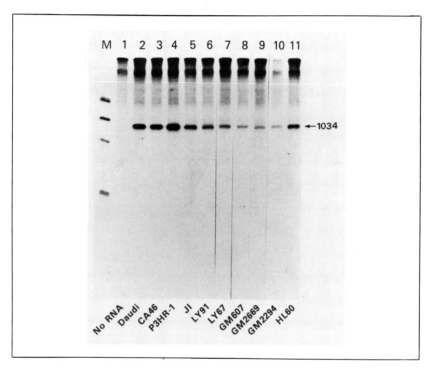

Fig. 2. Detection of the transcripts produced from the c-*myc* gene in various Burkitt lymphoma cells and human lymphoblastoid cell lines by S_1 nuclease analysis. The probe, cleaved with Bc1 1 and $5'$-^{32}p end labelled pRyc-7.4 plasmid, was heat denatured, hybridized in 80% formamide to 20 μg cytoplasmic RNA at 55 °C, digested with S_1 nuclease and analyzed by electrophoresis on a 7M urea 4% polyacrylamide gel [37]. RNA from Burkitt lymphoma cells. Lanes *2–7*, Daudi, CA46, P3HR-1, JI, LY91 and LY67. RNA from lymphoblastoid cell lines. Lanes *8–10*, GM607, GM2669 and GM2294. Lane *11*, RNA from promyelocytic leukemia cell line. HL-60. Lane *M*, size marker: ØX174 digested with HaeIII and $5'$-^{32}p-labelled. The human *myc* RNA protect a DNA fragment 1034 nucleotide long.

c-*myc* transcripts. These levels are higher than in the lymphoblastoid cells we have examined (figure 2) [15]. As shown in figure 2 the levels of c-*myc* transcripts of the three Burkitt lymphoma cell lines with the t(8;14) translocation (Daudi, CA46 and P3HR1) were even higher than in HL60 cells.

We have also examined the expression of the c-*myc* transcripts in Burkitt lymphomas with head-to-head (5' to 5') rearrangements of the c-*myc* gene with the Cμ gene by the Northern blotting procedure, us-

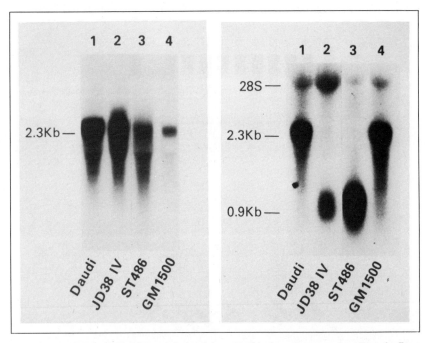

Fig. 3. Northern blotting analysis of three Burkitt lymphoma and an Epstein-Barr virus transformed lymphoblastoid cell line (GM1500). Cytoplasmic RNA was extracted [4] and 20 μg of RNA were added to each lane of 1% agarose gel. Following agarose gel electrophoresis and Southern transfer the nitrocellulose filters were hybridized *(A)* with the Ryc 7.4 probe (probe B) and *(B)* with the 5' exon probe (probe A). In lanes *1, 2* and *3* the RNAs from the lymphoma cell lines Daudi, JD38 IV and ST486 respectively. The DNA from GM1500 cells in lane *4*. All three lymphoma cell lines show a 2.3 kb c-*myc* transcript using the Ryc 7.4 probe. On the contrary we detect the 2.3 kb *myc* transcripts in Daudi and GM1500 cells using the 5' exon probe. We detected 0.9 kb transcripts hybridizing with the 5' exon probe in the two lymphoma cell lines with a rearranged c-*myc* oncogene. The intense hybridization of the 5' exon probe with 28 S ribosomal RNA is due to high G/C content [10].

ing nucleic acid probes specific for the untranslated leader of the *myc* message, (that is coded for by the first exon), and for the coding segment (second and third exon) of the c-*myc* oncogene [15]. As shown in figure 3B the probe specific for the first exon of the c-*myc* gene does not detect the 2.3 kb *myc* transcripts that are expressed in Burkitt lymphoma cells, in JD38 and ST486 cells, while the probe specific for the second and third exon does (figures 3A). This result indicates that JD38 and ST486 lymphoma cells with the t(8;14) translocation do not produce normal *myc* transcripts and that only the rearranged c-*myc* gene is transcribed, since we did not detect normal transcripts derived from the unrearranged c-*myc* gene on normal chromosome 8. In addition we detected short transcripts (~900 nucleotides long) derived from the first exon of the c-*myc* gene involved in the translocation (figure 3B) [16].

Transcription of the c-*myc* Oncogenes in Hybrids between Mouse Plasmacytoma and Burkitt Lymphoma Cells

We have also examined the expression of human and mouse c-*myc* transcripts in somatic cell hybrids between mouse plasmacytoma

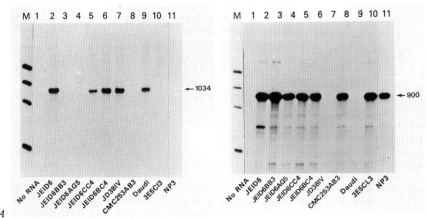

Fig. 4. S_1 nuclease analysis of c-*myc* RNAs in the hybrid cells between NP3 and Burkitt lymphoma cell lines with the t(8;14) chromosomal translocation. *(A)* Cytoplasmic RNA (20 μg) was hybridized with human c-*myc* probe or *(B)* mouse c-*myc* S_1 probe. The parental NP3 used for hybrid preparation is a non-producer mouse myeloma.

Table 1. Transcription of the mouse and human c-myc genes in mouse × human hybrids

Parental cells and hybrid clones	Human isozymes		Human chromosomes				Human oncogenes		Levels of c-myc transcripts	
	GSR	NP	8	$8q^-$	14	$14q^+$	c-mos	c-*myc*	Mouse	Human
P3HR-1 (BL)	+	+	+	+	+	+	+	+	−	+++
JE1D6 (NP3 × P3HR-1 hybrid)	+	+	+	+	−	+	+	++	+++	+++
BB3 (NP3 × P3HR-1 hybrid)	+	−	+	−	−	−	+	++	++++	−
AG5 (NP3 × P3HR-1 hybrid)	+	−	−	−	−	−	−	−	++++	−
CC4 (NP3 × P3HR-1 hybrid)	−	+	−	−	−	+	−	++	++++	+++
BC4 (NP3 × P3HR-1 hybrid)	−	+	+	+	−	−	−	+	++++	+++
NP3 (mouse plasma-cytoma)	−	−	−	−	−	−	−	−	−	−
JD38 (NBL)	+	+	+	+	+	+	+	+++	−	+++
253 A-B3 (NP3 × JD38)	+	−	−	−	+	−	+	−	+++	−
Daudi (BL)	+	−	+	+	+	+	+	+++	−	+++
3E5 CL3 (NP3 × Daudi hybrid)	+	+	+	+	+	−	+	+	+++	−

GSR, glutathione reductase, a marker of chromosome 8; NP, nucleoside phosphorylase, a marker of chromosome 14.

and Burkitt lymphoma cells by the S_1 nuclease protection procedure, using either a human or mouse c-*myc* cDNA probe [15]. As shown in figure 4 and table I only the hybrids with 14q$^+$ chromosome expressed human c-*myc* transcripts; hybrids with normal chromosome 8 did not (figure 4A and table I). All hybrids expressed high levels of mouse *myc* transcripts (figure 4B and table I). Therefore, we conclude that only the translocated *myc* gene is transcribed at a high level in Burkitt lymphoma cells and that the juxtaposition of the *myc* gene and of the heavy chain locus removes the translocated c-*myc* gene from its normal transcriptional control. Constitutive high level of expression of the c-*myc* gene product might be responsible for the expression of malignancy in Burkitt lymphoma.

Genetic Analysis of Burkitt Lymphoma Cells with the Variant t(8;22) and t(2;8) Chromosome Translocations

We have studied somatic cell hybrids between mouse myeloma cells and BL2 Burkitt lymphoma cells carrying a t(8;22) chromosome translocation for the presence (figure 5) and expression (figure 6) of human immunoglobulin chains and for c-*myc* oncogene. The results indicate that the c-*myc* oncogene remains on the 8q$^+$ chromosome, and that the excluded and rearranged Cμ allele translocates from chromosome 22 to this chromosome 8 (table II) [17]. As a result of the translocation, transcriptional activation of the c-*myc* oncogene on the

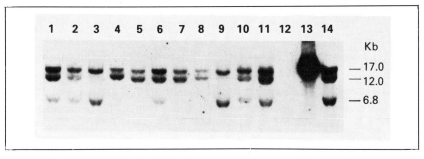

Fig. 5. Southern blotting analysis of NP3 × BL2 somatic cell hybrids following EcoR1 digestion of cellular DNA. The nitrocellulose filters were hybridized with a Cλ genomic clone. Lanes *1–11*, NP3 × BL2 hybrid DNAs. Hybrid 1–15 DNA is in lane *5;* hybrid 1–18 DNA is in lane *7* and hybrid 1–23 is in lane *8.* Hybrid 3–1 is in lane *3* and hybrid 1–9 DNA is in lane *9.* NP3 mouse myeloma DNA is in lane *12,* PAF DNA is in lane *13* and BL2 parental is in lane *14.*

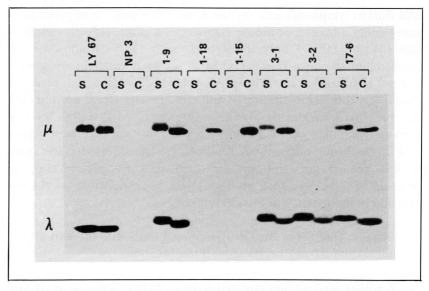

Fig. 6. Immunoprecipitation of human immunoglobulin chains produced by NP3 × BL2 hybrid clones. *S:* culture supernatant, *C:* cytosol. LY67 Burkitt lymphoma cells carry a t(8;22) translocation. Immunoprecipitation of BL2 culture supernatant or cytosol with the antihuman immunoglobulin antisera resulted in a band pattern identical to hybrids 1–9, 3–1 and 17–6 (data not shown).

rearranged chromosome 8 [8q$^+$] occurs (table II), while the c-*myc* oncogene on the normal chromosome 8 is transcriptionally silent [17]. These findings suggest that the translocation of a rearranged immunoglobulin λ locus to the 3' side of an unrearranged c-*myc* oncogene (figure 7) may enhance its transcription and contribute to malignant transformation [17].

We have also studied somatic cell hybrids between mouse myeloma and JI Burkitt lymphoma cells carrying a t(2;8) chromosome translocation for the expression of the human κ chains and for the presence and rearrangements of the human c-*myc* oncogene and the κ chain genes. Our results indicate that the c-*myc* oncogene is unrearranged and remains on the 8q$^+$ chromosome of JI cells (table III) [18]. Two rearranged Cκ genes were detected: the expressed allele on normal chromosome 2, and the excluded allele that was translocated from chromosome 2 to the involved chromosome 8 (8q$^+$) (table III). The distribution of Vκ and Cκ genes in hybrid clones retaining different human chromosomes indicated that Cκ is distal to the Vκ on 2p,

Table II. Ig genes and oncogenes in BL2 hybrids

Cell line	Human chromosomes[a]				Human Cλ genes			Expression of human λ chains	Human oncogenes		Transcripts of human c-myc
	8	8y+	22	22q-	17 kb (upper band)	12 kb (middle band)	6.8 kb (lower band)		c-myc	c-mos	
BL2	+	+	+	+	+	+	+	+	+	+	+++
NP3	−	−	−	−	−	−	−	−	−	−	−
1– 9	++	−	++	++	+++	−	+	+	+++	+++	n.d.
1–15	±	++	−	−	+++	+++	−	−	+++	−	+++
1–23	++	++	−	++	+++	−	−	−	+++	++	n.d.
3– 1	++	−	++	++	+++	−	+++	+++	+++	+++	−
3– 2	±	−	++	++	+++	++	+++	+++	+++	++	−
4–35	−	++	++	++	++	++	++	++	++	++	n.d.
17– 6	−	++	++	+	+	++	++	++	++	++	+++

[a] Frequency of metaphases with relevant chromosome
− = none; ± = <10%; + = 10–30%; ++ = >30%; n. d. = not done.

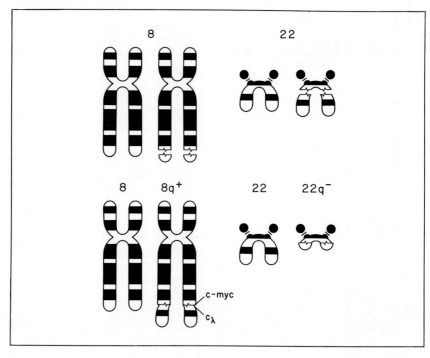

Fig. 7. Diagram of the t(8;22) translocation in Burkitt lymphoma. As shown in the figure the Cλ locus moves from its normal location on chromosome 22 to a region distal to the c-*myc* oncogene, which is untranslocated and unrearranged in BL2 cells.

and that the breakpoint in this Burkitt lymphoma is within the Vκ region (table III) [18]. High levels of transcripts of the c-*myc* gene were found when it resided on the $8q^+$ chromosome but not on the normal chromosome 8, demonstrating that translocation of a locus to a region distal to the c-*myc* oncogene enhances c-*myc* transcription (figure 8) [18].

Conclusions

The results described in this study indicate that in Burkitt lymphoma with the t(8;14) translocation the c-*myc* oncogene translocates from its normal position on band q24 of chromosome 14 to the heavy chain locus on chromosome 14. On the contrary, in the more infre-

Myc Activation in Burkitt Lymphoma   31

Table III. Human κ genes and oncogenes in JI × NP3 hybrids

Cells	Human chromosomes[a]				Human isozymes[b]				Human κ chains	Human genes		Human oncogenes	Human c-myc transcripts
										Cκ	Vκ		
JI	++	++	++	++	++	++	+	+	+	+	+	++	++
JI 4-5	–	++	–	–	++	++	–	–	–	++	–	++	+++
JI 4-5 B 7	–	+	–	–	+	–	–	–	–	+	–	++	++
JI 4-5 H 11	–	–	–	+	–	–	–	–	–	+	–	–	–
JI 5-4	+	–	–	–	–	–	–	–	+	–	+	+	–
JI 6-5	–	–	++	–	++	++	+++	–	+	+++	+	+	–
JI 4-2L	–	++	–	–	–	–	–	–	–	–	–	+	++
NP3	–	–	–	–	–	–	–	–	–	–	–	–	–

[a] Frequency of metaphases with relevant chromosomes
– = none; + = 10–30%; ++ = >30%.
[b] MDH = malate dehydrogenase; IDH = isocitrate dehydrogenase.

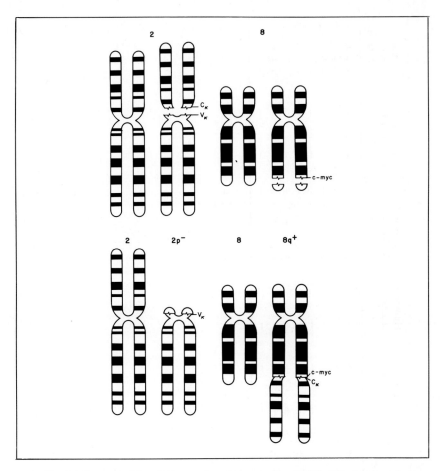

Fig. 8. Diagram of the t(2;8) translocation occurring in 5% of Burkitt lymphomas. As shown in the figure, the Vκ genes are proximal and the Cκ gene is distal on band p11 of chromosome 2. While some of the Vκ genes stay on the 2p⁻, the Cκ gene translocates to the involved chromosome 8 (8q⁺). The c-*myc* oncogene remains on the involved chromosome 8 (8q⁺).

quent Burkitt lymphomas with variant translocations the c-*myc* gene remains on chromosome 8 and either the λ or the κ locus translocates to a region distal to the c-*myc* oncogene. Independently of the head-to-head (5' to 5') or the head-to-tail (5' to 3') arrangement between the immunoglobulin and the c-*myc* loci the c-*myc* oncogene on the involved chromosome 8 becomes transcriptionally highly active and is removed from normal transcriptional control. Thus, high constitutive levels of c-*myc* expression occur as a result of the translocation in Burkitt lymphoma cells. These high constitutive levels of *myc* expression might have an essential role in the malignant proliferation of B cells in Burkitt lymphoma.

Summary

In Burkitt lymphomas with the 8;14 chromosome translocation we have shown that the c-*myc* oncogene translocates from its normal position on chromosome 8 to the heavy chain locus on chromosome 14. As a result, high levels of c-*myc* transcripts coded for by the translocated c-*myc* oncogene are expressed in the tumour cells, whereas the c-*myc* oncogene on normal chromosome 8 is transcriptionally silent.

We have also investigated Burkitt lymphoma cells with the t(2;8) and t(8;22) chromosome translocations by taking advantage of somatic cell genetic techniques to determine whether in the variant translocations similar mechanisms of c-*myc* activation occur. The results of these investigations indicate that in the various Burkitt lymphomas, the c-*myc* oncogene which is activated by the chromosomal translocation remains on chromosome 8 and that the genes for the constant regions of κ and λ light chains translocate from their normal position on chromosomes 2p and 22q respectively to a chromosomal region distal (3′) to the oncogene. These results indicate that activation of the c-*myc* gene can occur by placing the immunoglobulin gene sequences either in front (5′), or behind (3′) the oncogene.

References

1 Henle, W.; Henle, G.; Lennette, E. T.: Scient. Am. *241:* 48–59 (1979).
2 Manolov, G.; Manolova, Y.: Nature Lond. *237:* 33–36 (1972).
3 Zech, L.; Haglund, V.; Nilsson, N.; Klein, G.: Int. J. Cancer *17:* 47–56 (1976).
4 Miyoshi, I.; Hiraki, S.; Kimura, I.; Miyamato, K.; Sato, J.: Experientia *35:* 742–743 (1979).
5 Bernheim, A.; Berger, R.; Lenoir, G.: Cancer Genet. Cytogenet. *3:* 307–316 (1981).
6 Croce, C. M.; Shander, M.; Martinis, J.; Cicurel, L.; D'Ancona, G. G.; Dolby, T. W.; Koprowski, H.: Proc. natn. Acad. Sci. USA *76:* 3416–3419 (1979).
7 McBride, D. W.; Heiter, P. A.; Hollis, G. F.; Swan, D.; Otey, M. C.; Leder, P.: J. Exp. Med. *155:* 1680–1690 (1982).
8 Malcolm, S.; Barton, P.; Murphy, C.; Fergusson-Smith, M. A., Bentley, D. L.; Rabbitts, T. H.: Proc. natn. Acad. Sci. USA *79:* 4957–4961 (1982).
9 Erikson, J.; Martinis, J.; Croce, C. M.: Nature, Lond. *294:* 173–175 (1981).
10 Erikson, J.; Nowell, P. C.; Croce, C. M.: Proc. natn. Acad. Sci. USA *79:* 7824–7827 (1982).
11 Favera, R. dalla; Bregni, M.; Erikson, J.; Patterson, D.; Gallo, R. C.; Croce, C. M.: Proc. natn. Acad. Sci. USA *79:* 7824–7827 (1982).
12 Taub, R.; Kirsch, I.; Morton, C.; Lenoir, G.; Swan, D.; Tronick, S.; Aaronson, S.; Leder, P.: Proc. natn. Acad. Sci. USA *79:* 7837–7841 (1982).

13 Favera, R. dalla; Martinotti, S.; Gallo, R. C.; Erikson, J.; Croce, C. M.: Science 219: 963–967 (1983).
14 Favera, R. dalla; Wong-Staal, F.; Gallo, R. C.: Nature, Lond. 299: 61–63 (1982).
15 Nishikura, K.; ar-Rushdi, A.; Erikson, J.; Watt, R.; Rovera, G.; Croce, C. M.: Proc. natn. Acad. Sci. USA 80: 4822–4826 (1983).
16 ar-Rushdi, A.; Nishikura, K.; Erikson, J.; Watt, R.; Rovera, G.; Croce, C. M.: Science 222: 390–393 (1983).
17 Croce, C. M.; Thierfelder, W.; Nishikura, K.; Erikson, J.; Finan, J.; Lenoir, G.; Nowell, P. C.: Proc. natn. Acad. Sci. USA (in press, 1983).
18 Erikson, J.; Nishikura, K.; ar-Rushdi, A.; Finan, J.; Emanuel, B.; Lenoir, G.; Nowell, P. C.; Croce, C. M.: Proc. natn. Acad. Sci. USA (in press, 1983).

C. M. Croce, The Wistar Institute of Anatomy and Biology, 36th at Spruce Street, Philadelphia, PA 19104 (USA)

Molecular Mechanisms of Malignant Transformation by Oncornaviruses

K. Mölling, P. Donner, T. Bunte, I. Greiser-Wilke

Max-Planck-Institute for Molecular Genetics

Introduction

Oncornaviruses are vectors of oncogenes which are homologous to normal cellular genes and were picked up by the virus from the host cell during evolution. About one dozen of different oncogenes have been identified by now [3]. Their monoclonal cellular homologues can be detected throughout all animal species. The virus-associated oncogenes are not totally identical to their cellular homologues, they do not contain introns, furthermore they contain mutations and most importantly, their expression is regulated by the virus. Oncornaviruses are characterized by strong promoters which result in up-regulation of oncogene expression compared to the level of expression of the normal cellular genes (for review see [9]).

The recombinational event of oncornaviruses with cellular DNA which resulted in uptake of cellular genetic information into the viral genome occurred in many cases on the expense of viral replicating genes. The resulting viruses are therefore defective for replication and express their genetic information as covalently linked polyproteins consisting of a partially deleted structural gene, gag, and the oncogene. Monoclonal antibodies, directed against the gag portion of the polyproteins were used to characterize several gag-onc proteins [8]. One of them, gag-myc, belongs to the avian myelocytomatosis virus MC29, another one, gag-erbA, to the avian erythroblastosis virus, AEV, both of which are acute avian leukemia viruses [9]. A third protein, gag-fps, is coded for by the Fujinami sarcoma virus, FSV, which resembles the well-characterized Rous sarcoma virus, RSV [4]. The functions of these three gag-onc proteins were analyzed [1, 4, 12].

Materials and Methods

Isolation of the monoclonal antibody against p19 has been described [8]. Immunoaffinity column chromatography of gag-onc fusion proteins using the monoclonal antibody against p19 has been published previously [4].

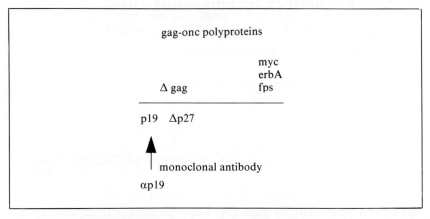

Fig. 1. Schematic drawing of gag-onc polyproteins, the transforming proteins of defective avian oncornaviruses. Monoclonal antibodies directed against p19, the N-terminal portion of gag-onc, were used to characterize the polyproteins. myc, erbA and fps indicate the oncogenes of MC29, AEV and FSV, respectively.

Results

To characterize various transforming proteins of oncornaviruses, we prepared monoclonal antibodies against the avian viral structural protein p19, which is the N-terminal portion of the gag-protein [8]. Figure 1 shows schematically how this monoclonal antibody allows identification of various transforming proteins due to their property of being gag-onc polyproteins. To elucidate the function of these proteins during malignant transformation, their cellular localizations were analyzed by using the monoclonal antibodies for indirect immunofluorescence. The second antibody was labelled with fluoresceine. The cells analyzed were transformed fibroblasts which expressed the gag-onc polyproteins but no other viral genes since they were so-called non-producer cells. Besides the three viruses MC29, AEV and FSV, a deletion mutant of MC29, Q10H, was analyzed which has a deletion in the myc portion [15] and two other members

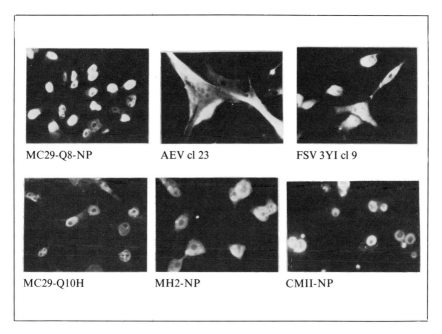

Fig. 2. Indirect immunofluorescence of non-producer fibroblasts transformed by various avian oncornaviruses using monoclonal antibodies against p19. MC29-Q8-NP is an established quail fibroblast non-producer cell line transformed by myelocytomatosis virus; AEV cl 23 is a clone of avian erythroblastosis virus transformed chicken non-producer fibroblasts; FSV-3Y1-cl 9 represents a non-producer rat-cell line transformed by Fujinami sarcoma virus [6]; MC29-Q10H is a non-producer quail fibroblastic cell line infected with a deletion mutant of MC29 [15]; MH2-NP and CMII-NP are non-producer quail cells transformed by Mill-Hill virus and myelocytomatosis virus, respectively.

of the MC29 family, MH2 and CMII (figure 2). The gag-myc protein is located in the nuclei of MC29-, Q10H- and CMII-transformed cells. MH2 is an exception. Together with AEV and FSV it gives rise to total cell fluorescence.

Furthermore, all these cells were metabolically labelled and their polyproteins extracted by immunoaffinity column chromatography which results in about 3000-fold purification in a single step. The purified ^{35}S-methionine-labelled proteins were analyzed on polyacrylamide gels and are shown in figure 3. Their molecular weights are indicated at the sides and expressed in kilodaltons.

Since the most well-known transforming protein pp60^{src} of Rous sarcoma virus has already been identified as a protein kinase [5], all

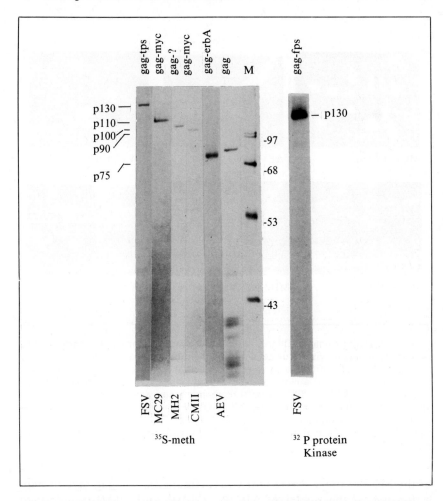

Fig. 3. Purified transforming gag-related proteins. Transformed non-producer cells ($\sim 10^8$ cells) shown in figure 1 were metabolically labelled for 4 h with 250 µCi/ml of ^{35}S-methionine and the transforming proteins isolated by immunoaffinity column chromatography using monoclonal antibodies against p19 [4]. The purified proteins were analyzed on 10% polyacrylamide slab gels and exposed for autoradiography. M indicates molecular weight markers. The numbers indicate molecular weights in kilodaltons. The protein kinase assay with purified p130gag-fps protein was performed under standard protein kinase assay condition [6]. The reaction mixture was then analyzed by polyacrylamide gel electrophoresis and autoradiography.

the isolated transforming proteins shown in figure 3 were tested for protein kinase activity. Only the purified p130gag-fps protein exhib-

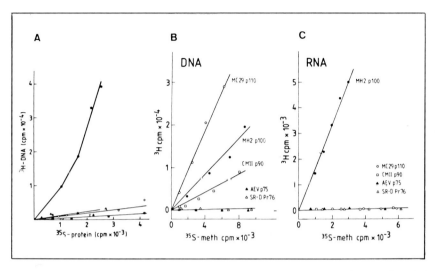

Fig. 4. Filter-binding assays for DNA- or RNA-protein interaction studies. *(A)* Purified gag-myc proteins of MC29 wildtype (●) and three deletion mutants (o, ▲, ■) were tested for their DNA-binding abilities in a filter-binding assay in the presence of 50 mM NaCl as described [4]. ^{3}H-DNA consisted of double-stranded cellular DNA of about 1 kb in length (10^5 cpm/µg). *(B)* Purified gag-onc proteins as indicated were tested for interaction with DNA in vitro. *(C)* Analysis of RNA-protein interaction with gag-onc proteins in vitro. ^{3}H-RNA consisted of single-stranded cellular ribosomal RNA (10^5 cpm/µg). Specific activities of all the proteins were about 8×10^4 cpm ^{35}S-meth/µg of protein.

ited protein kinase activity and strongly phosphorylated itself (figure 3, right). The others were all negative.

To further characterize functions of the transforming proteins, their cellular localizations were analyzed further. MC29-transformed fibroblasts were fractionated to isolate chromatin. A portion of the gag-myc protein was found to be associated with the chromatin [1]. This result led to the analysis of DNA-protein interaction in vitro. The p110gag-myc protein of MC29 bound to double-stranded DNA in vitro (figure 4 A). The DNA-binding ability correlated with transformation, since the transforming proteins of three deletion mutants derived from CM29 wild-type exhibited a reduced DNA-binding ability (figure 4 A). How the DNA-protein interaction results in malignant transformation is still unknown. Several mechanisms can be envisioned, e. g. by methylation or inhibition of methylation of DNA, by binding to promoters or other regulatory regions of the DNA and modification of gene expression, by blocking of transcription etc. The gag-myc protein does not appear to be a methylase (*Bunte et al.,* unpublished observation); all the other mechanisms need to be investigated. Even though the gag-myc protein of the deletion mutants show reduced ability to bind to DNA, they are still nuclear antigens (see Q10H, figure 2). This observation needs to be taken into consideration in the search for the biological role of the gag-myc protein.

Two MC29 family members, MH2 and CMII, were also tested for DNA-binding ability of their gag-onc polyproteins. CMII is highly related to MC29 and codes for a gag-myc protein which is located in the nucleus and binds to DNA in vitro (figures 2 and 4 B). In contrast, MH2, which also belongs to the MC29 family, codes for a gag-onc protein which is cytoplasmic (figure 2) and binds to RNA in vitro (figure 4). This is a novel property for a gag-onc protein and not typical of a gag-myc type protein [2]. MH2, therefore, probably codes for two onc-proteins, with only one of them fused to gag. The one not fused to gag is most likely the myc-specific one [11, 13] which will only become detectable when myc-specific sera become available. Why some viruses pick up more than one oncogene is unclear. Whether both of them are involved in the process of malignant transformation or only one of them is also not yet known. The MH2-specific gag-onc fusion protein appears to inhibit in vitro translation according to preliminary evidence. Inhibition of some translational processes might also interfere with differentiation of infected cells. In the case of AEV which also codes for two oncogenes, one of them, gag-erbA, may not be directly involved in malignant transformation but inhibition of differentiation [7].

The gag-erbA protein has proven to be the most difficult protein to characterize. The only information about it comes from immunofluorescence (figure 2) and cellular fractionation studies [4]. It is a cytoplasmic protein but again it differs from the MH2 gag-onc protein in that it does not bind to DNA or RNA in vitro [4] (figure 4 B and C). The erbB protein has recently been characterized as a membrane-associated glycoprotein [14].

The gag-fps protein of FSV which was shown to be a protein kinase after purification (figure 3). (*Donner et al.,* unpublished results) was further analyzed for its biological role during transformation by in vitro phosphorylation of various substrates. One of the potential target proteins of the RSV-specific protein kinase has been the

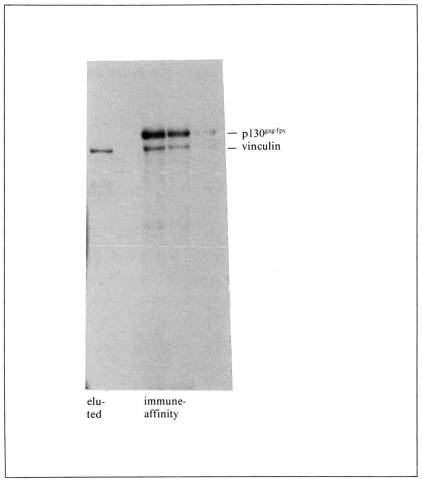

Fig. 5. Vinculin phosphorylation in vitro by FSV-protein kinase. p130gag-fps immobilized on the immunoaffinity column was allowed to phosphorylate vinculin. The phosphorylated vinculin was then eluted and analyzed on a polyacrylamide gel and by autoradiography (left: *eluted*). The remaining material was removed from the column with low-pH-buffer to disrupt the antigen-antibody complex [4]. p130 as well as residual vinculin which were both phosphorylated on the column were washed off together by this procedure and analyzed on a gel and by autoradiography. Three subsequent fractions of the washing procedure are shown (right: *immuneaffinity*).

cytoskeletal protein vinculin [16]. Therefore, purified vinculin was also tested with the p130gag-fps protein kinase. Indeed, the p130gag-fps protein kinase phosphorylated vinculin in vitro (figure 5). To the left of figure 5 an experiment is shown in which vinculin was phosphorylated on the p130gag-fps protein kinase immobilized on the immunoaffinity column and eluted after phosphorylation. Subsequently the p130gag-fps protein and residual amounts of vinculin were eluted from the immunoaffinity column after breaking the immune complex. Since p130gag-fps phosphorylated itself, both proteins are detectable as phosphoproteins (figure 5 right, three different aliquots solubilized from the immunoaffinity column). Phosphorylation of vinculin and the p130gag-fps protein occurred exclusively in tyrosine (not shown) – a specificity known for protein kinases from oncornaviruses [10].

Discussion

Table I summarizes the data described on transformation specific oncornaviral polyproteins. The molecular mechanisms involved in transformation which are caused by oncornavirus-specific oncogenes are obviously not uniform. They range from nuclear antigens to membrane proteins and involve interaction with DNA or modification of membrane structural components. Yet another transformation mechanism exists for those oncornaviruses which do not carry their own oncogenes. These viruses act indirectly. It will be interesting to elucidate whether the phenomena observed represent various stages of a multistage process resulting in malignant transformation or whether they all represent independent transforming mechanisms.

Table I. Summary of transformation-specific proteins of several avian oncornaviruses

Virus	Name	Oncogene	Disease in animal	Molecular mechanism of malignant transformation
I Sarcoma virus	RSV FSV	src fps	Sarcomas	tyr-specific protein kinase
II Acute leukemia virus	MC29 family AEV E26	myc erbA, B myb-ets	Acute leukemias (sarcomas)	DNA-binding protein A: ? B: glycoprotein DNA-binding protein-E26
III Lymphatic leukemia virus	RAV		Lymphomas	indirect

References

1. Bunte, T.; Greiser-Wilke, I.; Donner, P.; Moelling, K.: Association of gag-myc proteins from avian myelocytomatosis virus wild-type and mutants with chromatin. EMBO J. *8;* 918–927 (1982).
2. Bunte, T.; Greiser-Wilke, I.; Moelling, K.: The transforming protein of the MC29-related virus CMII is a nuclear binding protein whereas MH2 codes for a cytoplasmic RNA-DNA binding protein. EMBO J. *7;* 1087–1092 (1983).
3. Coffin, J. M. et al.: A proposal for naming host cell derived inserts in retrovirus genomes. J. Virol. *40;* 953–957 (1981).
4. Donner, P.; Greiser-Wilke, I.; Moelling, K.: Nuclear localization and DNA-binding of the transforming gene product of avian myelocytomatosis virus. Nature *296;* 262–266 (1982).
5. Erikson, R. L.: The transforming protein of avian sarcoma viruses and its homologue in normal cells. Curr. Top. Microbiol. Immunol. *91;* 25–40 (1981).
6. Feldmann, R. A.; Hanafusa, T.; Hanafusa, H.: Characterization of protein kinase activity associated with the transforming gene product of Fujinami sarcoma virus. Cell *22;* 757–765 (1980).
7. Frykberg, L.; Palmieri, S.; Beug, H.; Graf, T.; Hayman, M. J.; Fennström, B.: Transforming capacities of avian erythroblastosis virus mutants deleted in the erbA and erbB oncogenes. Cell *32;* 227–238 (1983).
8. Greiser-Wilke, I.; Owada, M. K.; Moelling, K.: Isolation of monoclonal antibodies against the avian oncornaviral protein p19. J. Virol. *39;* 325–329 (1981).
9. Hayward, W. S.; Neel, B. G.: Retroviral gene expression. Curr. Top. Microbiol. Immunol. *91;* 218–244 (1981).
10. Hunter, T.; Sefton, B. M.: The transforming gene product of Rous sarcoma virus phosphorylates tyrosine. Proc. natn. Acad. Sci. USA *77;* 1311–1315 (1980).
11. Linial, M.: Two retroviruses with similar transforming genes exhibit differences in transforming potential. Virology *119;* 382–391 (1982).
12. Moelling, K.; Owada, M. K.; Greiser-Wilke, I.; Bunte, T.; Donner, P.: Biochemical characterization of transformation-specific proteins of acute avian leukemia and sarcoma viruses. J. Cell. Biochem. *20;* 63–69 (1982).
13. Pachl, C.; Biegalk, B.; Linial, M.: RNA and protein encoded by MH2 virus: evidence for subgenomic expression of v-myc. J. Virol. *45;* 133–139 (1983).
14. Privalsky, M. L.; Sealy, L.; Bishop, J. M.; McGrath, J. P.; Levinson, A. D.: The product of the avian erythroblastosis virus erbB locus is a glycoprotein. Cell *32:* 1257–1267 (1983).
15. Ramsay, G.; Graf, T.; Haymann, M. J.: Mutants of avian myelocytomatosis virus with smaller gag gene-related proteins have an altered transforming ability. Nature *288;* 170–172 (1980).
16. Sefton, B. M.; Hunter, T.; Ball, E. H.; Singer, S. J.: Vinculin: A cytoskeletal target of the transforming protein of Rous sarcoma virus. Cell *24:* 165–174 (1981).

Priv.-Doz. Dr. Karin Mölling, Max-Planck-Institut für Molekulare Genetik, Ihnestraße 63–73, D-1000 Berlin 33

Cancer-Associated Carbohydrate Antigens Detected by Monoclonal Antibodies

V. Ginsburg, P. Fredman, J. L. Magnani

National Institute of Arthritis, Diabetes, and Digestive and Kidney Diseases, National Institutes of Health, Bethesda, Md., USA

The carbohydrates on cell surfaces change during development as they are probably involved in cell recognition. It is these developmentally regulated changes that allow some antibodies directed against carbohydrates to discriminate among various tissues, both normal and malignant. The most important glycosyl residues as far as immunological specificity is concerned are generally at non-reducing ends of the carbohydrate chains. These "immunodominant" residues determine the antigenic specificity of the molecules in which they occur. Because the terminal sequences of sugars in the sugar chains of glycoproteins and glycolipids are sometimes identical, some antibodies directed against carbohydrates react with both glycoproteins and glycolipids.

Unlike protein antigens, which are primary gene products, carbohydrate antigens are secondary gene products [1, 2]. The primary gene products are glycosyltransferases which add single glycosyl residues to the growing chains, and it is the presence or absence of these enzymes that determine which particular chains or antigens are synthesized. Over 70 different disaccharide sequences occur in the complex carbohydrates of cell surfaces, most of which are products of separate glycosyltransferases. As the acceptor specificity of these enzymes is greater than just single glycosyl residues, the total number of glycosyltransferases that a cell can potentially produce is probably in the hundreds. The expression of these enzymes varies during development, and it is this variation that accounts for the characteristic pattern of carbohydrate structures that occur in different tissues and for the appearance and disappearance of carbohydrate antigens. In addition, because some glycosyltransferases compete for the same chains, the amount of antigen formed may depend on the relative activities of those enzymes that are present [3]. Thus, there are several ways that carbohydrate antigens are modulated as illustrated in figure 1. Assuming the trisaccharide sequence labelled B is an antigen, the level of antigen *B* will be maximal if only enzymes *a* and *b* are active. Its level will be reduced if enzymes *c* or *d* are active, and it will disappear if enzymes *a* or *b* disappear.

In attempts to obtain monoclonal antibodies specific for various cancers, mice and rats have been immunized with human tissues in many laboratories. Some of the antibodies derived from spleen cells of the immunized animals have an apparent specificity for certain cancers and are directed against carbohydrates. We have tested about 325 monoclonal antibodies for carbohydrate specificity, sent to us from various laboratories in the past 3 years ([4–9] and unpublished data). Of these antibodies, about one third are directed against carbohydrates as determined by solid-phase radioimmunoassay and autoradiography (table I). The first five antigens are associated with the human ABO and Lewis blood group systems. Of these, the antibody against the H type 1 antigen is especially interesting as it specifically

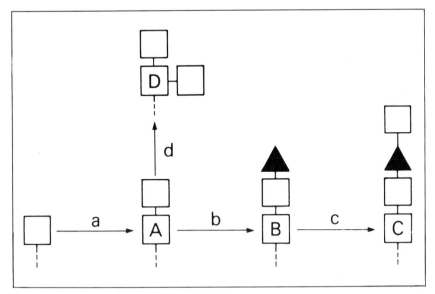

Fig. 1. Modulation of expression of carbohydrate antigens. *A, B, C* and *D* are different carbohydrate antigens; *a, b, c* and *d* are glycosyltransferases (see text).

precipitates the epidermal growth factor receptor of the human epidermoid carcinoma cell line A431 (a glycoprotein of M_r 170,000) and also reacts with glycolipids containing the H type 1 sequence of sugars [9].

About one sixth of the antibodies tested are directed against a sugar sequence found in the human milk oligosaccharide lacto-N-fucopentaose III [10] and which must be extremely immunogenic in mice and rats. A glycolipid containing lacto-N-fucopentaose III was first isolated from a human adenocarcinoma [11]. The same sugar sequence minus the glucosyl residue occurs in higher glycolipids [12] and also in glycoproteins [13]. Antibodies against this sugar sequence detect a stage-specific embryonal antigen (called SSEA-1) of the murine embryo and teratocarcinoma [12, 14]; they also detect an antigen (called My-1) that is strongly expressed in human granulocytes and granulocyte precursor cells but not in normal peripheral blood lymphocytes, monocytes, platelets or red cells [8]; and they also detect an antigen characteristic of human small cell carcinoma, adenocarcinoma and squamous cell carcinoma of the lung [7]. Interestingly, the My-1 antigen transiently appears on My-1-negative mouse fibroblasts following transfer of DNA from human myeloblastic of lymphoblastic leukemia cells [15]. As My-1 is a carbohydrate antigen, DNA transfer presumably alters the pattern of glycosyltransferases normally present in the recipient cells, and it is these altered glycosyltransferases that catalyze the synthesis of My-1.

The antibodies directed against the sialylated Le^a antigen [6] may be useful diagnostically as they detect antigen in the serum of most patients with gastrointestinal and pancreatic cancer [16]. Although the sialylated Le^a sequence of sugars occurs in the gangliosides of pancreatic and gastrointestinal cancers [6, 17], the antigen detected in the sera of patients are mucins (high molecular weight carbohydrate-rich glycoproteins). The evidence for this is as follows [18]: little antigen is extracted by organic solvents from sera and that which is extracted remains at the origin under conditions of thin-layer chromatography where the ganglioside antigen migrates up the plate. Upon gel filtration of serum on Sephacryl S-400, the antigen elutes in the void volume, indicating an $M_r \geq 5 \times 10^6$. Incubation for 5 h at 37 ° in 0.1 N NaOH destroys the serum antigen but does not affect the ganglioside antigen. The density of the serum antigen as determined in a CsCl gradient is 1.50 g/ml, while in 4M guanidine·HCl its density is

1.43 g/ml. Finally, antigen affinity-purified by anti-sialylated Le^a antibody from the serum of a cancer patient belonging to the Le (a−b+) blood group contains Le^b antigen, consistent with the multiple antigenic specificities exhibited by mucins. The occurrence of mucins in the blood of cancer patients has been reported many times [19].

In a previous study [6], sialylated Le^a antigen was not detected by solid-phase radioimmunoassay in extracts from normal adult tissues. By immunoperoxidase labelling of normal tissue sections, however, the antigen was found in a layer of ductal cells in normal pancreas and a layer of cells in normal salivary glands and bronchial epithelium that secrete mucins [20] and by autoradiography low levels of ganglioside antigen were detected in extracts of normal pancreas [17]. The antigen is also found in salivary mucins from most normal indi-

Table I. Monoclonal antibodies directed against carbohydrates

No. of antibodies (out of 325 tested)	Antigen	Structure
1	H type 1	Fucα1-2Galβ1-3GlcNAc...
1	H type 2	Fucα1-2Galβ1-4GlcNAc...
5	H type ?	Fucα1-2Galβ1-?GlcNAc...

4	Le^b	Fucα1-2Galβ1-3GlcNAc...
		                       4
		                       |
		                      Fucα1

3	B	Fucα1-2Galβ1-4GlcNAc...
		                       3
		                       |
		                      Galα1

55	Lacto-*N*-fucopentaose III (also called SSEA-1, My-1, Le^x and X-hapten)	Galβ1-4GlcNAc...
		             3
		             |
		            Fucα1

2	Sialylated Le^a	NeuNAcα2-3Galβ1-3GlcNAc...
		                             4
		                             |
		                          Fucα1

26	Unidentified carbohydrate sequences in glycolipids and/or glycoproteins	
97 total		

viduals belonging to the Le (a+b−) or Le (a−b+) blood group and is not found in salivary mucins from normal individuals belonging to the Le (a−b−) blood group [21]. About 7% of the population belong to the Le (a−b−) blood group because they lack the fucosyltransferase that catalyzes the synthesis of the sugar sequence Fucα1-4GlcNAc ... [22]. As a consequence, cancer patients belonging to the Le (a−b−) blood group cannot synthesize sialylated Le^a antigen [23]. Since the carbohydrate composition of mucins varies in different patients, monoclonal antibodies directed against other carbohydrate sequences found in mucins might be used together with the anti-sialylated Le^a antibody for a more sensitive serum test for gastrointestinal and pancreatic cancer.

The monosaccharide residues commonly found in the complex carbohydrates of cell surfaces include three hexoses (*D*-glucose, *D*-galactose and *D*-mannose), one 6-deoxyhexose (*L*-fucose), two hexuronic acids (*D*-glucuronic acid and *L*-iduronic acid) one pentose (*D*-xylose), and three amino sugars (*N*-acetyl-*D*-glucosamine, *N*-acetyl-*D*-galactosamine and *N*-acetylneuraminic acid). Fucose and *N*-acetylneuraminic acid are usually found only at the nonreducing ends of sugar chains, which might partially explain their importance as antigens (see table I). The curious distribution of glucose supports the idea that complex carbohydrates function in cell recognition [24]. This hexose is the most abundant sugar found in nature and occurs in many complex carbohydrates of plants and bacteria. In animals it comprises the reserve polysaccharide glycogen and substantial amounts are found free in body fluids. Yet with rare exceptions, chiefly collagen-like structural proteins, it is not found in mammalian glycoproteins. In glycolipids it occurs as the glycosyl residue closest to the lipid, and not in the distal parts of the sugar chains which extend outwards. The exclusion of glucose has a rational basis if the carbohydrate structures of cell surface were indeed binding complementary molecules, presumably mammalian lectins [25]: the efficiency of binding sites based on glucosyl residues would be impaired by the free glucose of body fluids, much as haptens inhibit antigen-antibody reactions. Evolutionary selection against the impairment would exclude glucose as a component of these surfaces.

Summary

To obtain monoclonal antibodies specific for various cancers, mice and rats have been immunized with human tissues in many laboratories. Some of the antibodies derived from spleen cells of the immunized animals have a apparent specificity for certain cancers and are directed against carbohydrates. In the past 3 years we have tested about 325 antibodies, sent to us from various laboratories for carbohydrate specificity. Of these antibodies, 97 are directed against carbohydrates as determined by solid-phase radioimmunoassay and autoradiography. Of the 97 antibodies, 7 are directed against the H type 1 or H type 2 antigens, 4 against the Le^b antigen, 3 against the B antigen, 55 against a sugar sequence found in the human milk oligosac-charide lacto-N-fucopentaose III, 2 against a sialylated Le^a antigen and 26 against unidentified carbohydrate sequences in glycolipids and/or glycoproteins.

References

1 Ginsburg, V.: in Meister (ed.), Advances in Enzymology, vol. 36, p. 131 (John Wiley and Sons, New York 1972).
2 Watkins, W. M.: in Harris, Hirschhorn, (eds.), Advances in Human Genetics, vol. 10, p. 1. (Plenum, New York 1980).
3 Rohr, T. E.; Smith, D. F.; Zopf, D. A.; Ginsburg, V.: Archs Biochem. Biophys. 199, 265–269 (1980).
4 Brockhaus, M.; Magnani, J. L.; Blaszczyk, M.; Steplewski, Z.; Koprowski, H.; Karlsson, K. A.; Larsson, G.; Ginsburg, V.: J. biol. Chem. 256: 13223–13225 (1981).
5 Brockhaus, M.; Magnani, J. L.; Herlyn, M.; Blaszczyk, M.; Steplewski, Z.; Koprowski, H.; Ginsburg, V.: Archs Biochem. Biophys. 217: 647–751 (1982).
6 Magnani, J. L.; Nilsson, B.; Brockhaus, M.; Zopf, D.; Steplewski, Z.; Koprowski, H.; Ginsburg, V.: J. biol. Chem. 256: 14365–14369 (1982).
7 Huang, L. C.; Brockhaus, M.; Magnani, J. L.; Cuttitta, F.; Rosen, S.; Minna, J. D.; Ginsburg, V.: Archs Biochem. Biophys. 229: 318–320 (1983).
8 Huang, L. C.; Civin, C. I.; Magnani, J. L.; Shaper, J. H.; Ginsburg, V.: Blood 61: 1020–1023 (1983).
9 Fredman, P.; Richert, N. D.; Magnani, J. L.; Willingham, M. C.; Pastan, I.; Ginsburg, V.: Fed. Proc. 42: 1352 (1983); J. biol. Chem. 258: 11206–11210 (1983).
10 Kobata, A.; Ginsburg, V.: J. biol. Chem. 244: 5496–5502 (1969).
11 Yang, H.-J.; Hakomori, S.: J. biol. Chem. 246: 1192–1200 (1971).
12 Hakomori, S.; Nudelmann, E.; Levery, S.; Solter, D.; Knowles, B. B.: Biochem. biophys. Res. Commun. 100: 1578–1586 (1981).
13 Lloyd, K. O.; Kabat, E. A.; Licerio, E.: Biochemistry 7: 2276–2990 (1968).
14 Gooi, H. C.; Feizi, T.; Kapadia, A.; Knowles, B. B.; Solter, D.; Evans, M. J.: Nature 292: 156–158 (1981).
15 Chang, L. J.-A.; Gamble, C. L.; Izaguirre, C. A.; Minden, M. D.; Mak, T. W.; McCulloch, E. A.: Proc. natn. Acad. Sci. USA 79: 146–150 (1982).
16 Herlyn, M.; Sears, H. F.; Steplewski, A.; Koprowski, H.: J. clin. Immunol. 2: 135–140 (1982).
17 Hansson, G. C.; Karlsson, K. A.; Larsson, G.; McKibbin, J. M.; Blaszczyk, M.; Herlyn, M.; Steplewski, Z.; Koprowski, H.: J. biol. Chem. 258: 4091–4099 (1983).

18 Magnani, J. L.; Steplewski, A.; Koprowski, H.; Ginsburg, V.: Cancer Res. (in press, 1983).
19 Race, R. R.; Sanger, R.: Blood Groups in Man; 6th ed., p. 40 (Blackwell, Oxford 1975).
20 Atkinson, B. F.; Ernst, C. S.; Herlyn, M.; Steplewski, A.; Sears, H. F.; Koprowski, H.: Cancer Res. *42:* 4820–4823.
21 Brockhaus, M.; Wysocka, M.: Unpublished data.
22 Grollmann, E. F.; Kobata, A.; Ginsburg, V.: J. clin. Invest. *48:* 1489–1494 (1969).
23 Koprowski, H.; Blaszczyk, M.; Steplewski, A.; Brockhaus, M.; Magnani, J. L.; Ginsburg, V.: Lancet *i:* 1332–1333 (1982).
24 Gesner, B. M.; Ginsburg, V.: Proc. natn. Acad. Sci. (USA) *52:* 750–755 (1964).
25 Barondes, S. H.: Annu. Rev. Biochem. *50:* 207–231 (1981).

Victor Ginsburg, M. D., National Institute of Arthritis, Diabetes, and Digestive and Kidney Diseases, National Institutes of Health, Bethesda, MD 20205 (USA)

Monoclonal Antibodies Reveal Saccharide Structures of Glycoproteins and Glycolipids as Differentiation and Tumour-Associated Antigens

T. Feizi

Applied Immunochemistry Research Group, Clinical Research Centre, Harrow, UK

Introduction

The last several years have witnessed a resurgence of interest in the identification of antigenic markers which distinguish tumour cells from their normal counterparts, foetal cells from those of adults and one differentiated cell from another. This renewed interest is due to the introduction of the hybridoma technique [1] which has made it possible to produce monoclonal antibodies of desired specificities and to identify with precision the antigens that they recognise. It has been hoped that such "tailor-made" monoclonal antibodies would single out antigenic markers which would be useful for the diagnosis of cancer, the immunotherapy of tumours and the typing of leukemic cells and not least for detecting important components which change during successive stages of development and differentiation.

In this report, I shall review our studies with natural-monoclonal and hybridoma-derived antibodies which have shown that a number of antigens behaving as tumour-associated or differentiation antigens of man and mouse are carbohydrate structures rather than proteins. They belong to a family of carbohydrate structures which also includes the major blood group antigens A, B, H, Le[a] and Le[b]. They occur both on glycoproteins and glycolipids of the cell surface [2, 3] and none of them is uniquely specific for tumours or individual cell types.

The Three Carbohydrate Domains on Blood Group-Related Oligosaccharides

The blood group-related carbohydrate chains consist of three domains [4] *core, backbone* and *peripheral* (figure 1). Monosaccharides in the core region vary according to whether the oligosaccharides are joined by N-or O-glycosidic linkage to protein [4, 5] or to lipid [6].

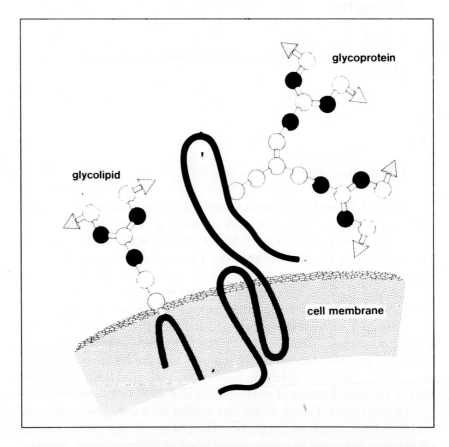

Fig. 1. Schematic presentation of a cell membrane showing carbohydrate chains on a glycoprotein and a glycolipid projecting from the surface. Carbohydrate chains consist of monosaccharides in three domains; "core", ☉, "backbone", ●, ○ and "peripheral", △, linked in various ways. The galactose and N-acetylglucosamine residues shown in table I occur as part of the backbone regions; monosaccharides in the peripheral regions may consist of fucose, galactose, N-acetylgalactosamine and sialic acid residues.

The antigens I shall discuss in this report are found associated with the backbones and the peripheral regions. The backbone regions consist of alternating galactose (Gal) and N-acetylglucosamine (GlcNAc) residues joined by two types of linkage with differing antigenicities, known as type 1 and type 2 chains [7] as follows:

Galβ1-3GlcNAc Type 1
Galβ1-4GlcNAc Type 2

in linear sequence the repeating disaccharide units are joined to one another by 1-3 linkage (table I, figures 2 and 3); branch points are formed by disaccharide units joined by 1-6 linkage to galactose or N-acetylgalactosamine residues in the backbone or core regions re-

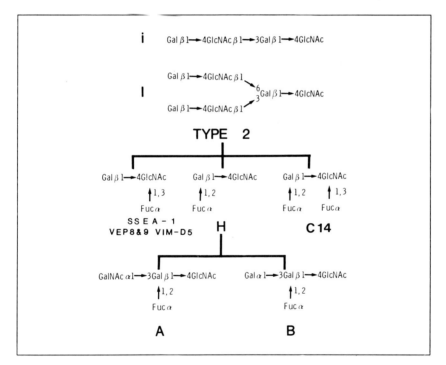

Fig. 2. Interrelationships of the i and I antigens expressed on the linear and branched type 2 sequences, the SSEA-1, VEP8/9 and VIM-D5 expressed on α1-3 fucosylated type 2 chains, C 14 expressed on difucosylated type 2 chains and the *H, A* and *B* antigens based on type 2 backbone structures. The branched backbone sequence expresses I antigenic determinants recognised by various anti-I antibodies including anti-I Ma, and the hybridoma antibodies M 39 and M 18. The latter determinant also occurs as a branch attached to type 1 backbones as in figure 3.

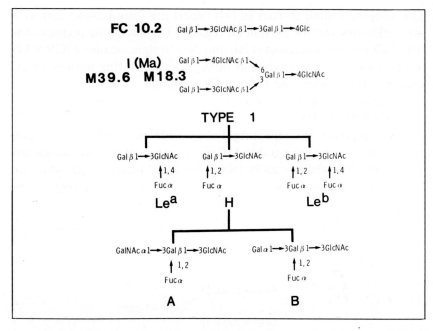

Fig. 3. Interrelationships of the FC 10.2 determinant expressed on type 1 based backbone structures, the I(Ma) determinant forming a type 2 branch point on type 1 based backbone and the Le^a, Le^b antigens and the *H*, *A* and *B* antigens based on type 1 backbone structures.

spectively. Backbone regions are extremely variable [4] and may be longer or shorter than shown in figure 1 and they may be linear or branched in different cell types; their antigenicity varies accordingly. The backbone sequences may be further glycosylated with monosaccharides such as fucose (Fuc), galactose or N-acetylgalactosamine (GalNAc) to give rise to peripheral regions with differing antigenicities. Among these the best known are the major blood group antigens, A, B, H, Le^a and Le^b [7]. The peripheral glycosylations often mask the antigenicities of the underlying backbone structures [2, 8].

It is of considerable interest that there occur both natural and hybridoma-derived antibodies against this family of saccharide structures. Certain of these structures behave as tumour-associated antigens in some individuals but not in others [9]. Furthermore, certain of these antigens are normal cellular components of some tissues but they behave as tumour-associated antigens in others, as discussed be-

Table 1. Structures behaving as differentiation or tumour-associated markers in man and mouse

Markers in Man	Antigens Designation[a]	Structure	Markers in Mouse
Distinctive antigen of *foetal erythrocytes* [8, 10]	i	Galβ1-4GlcNAcβ1-3Galβ1-4GlcNAc	Marker of *embryonic endoderm* [15]
Distinctive antigen of *erythrocytes of adults* [8, 10]	I	Galβ1-4GlcNAcβ1 ⧹6 Galβ1-4GlcNAc Galβ1-4GlcNAcβ1 ⁄3	Marker of *earliest embryo cells* [15]
Distinctive antigen of *normal stomach mucosa* in 'non-secretors' [11] Tumour-associated antigen in *stomach* of 'secretors' [11] Differentiation antigen of *normal breast epithelium* [12]	I(Ma), M39, M18	Galβ1-4GlcNAcβ1 ⧹6 Gal Galβ1-4GlcNAcβ1 ⁄	

Distinctive marker of *granulocytes* among cells of the peripheral blood [19, 20]	SSEA-1, VEP8/9 VIM-D5	Galβ1-4GlcNAc
		      |1,3
		     Fucα
			Marker of *eight cell stage embryo* [16, 17]

Marker of *carcinoma of colon* intra-cellular antigen of *granulocytes* [24]	C14	Galβ1-4GlcNAc
		  |1,2 |1,3
		  Fucα Fucα	

Marker of *embryonic endoderm* [26]	FC10.2	Galβ1-3GlcNAcβ1-3Galβ1-4Glc/GlcNAc	

[a] The i, I and I(Ma) antigens are recognized by natural-monoclonal autoantibodies; the other antigens are recognized by hybridoma-derived antibodies.

low. The unexpected aspect of all these findings (and this is supported by observations from other groups) is that no unique structures have been revealed thus far.

Tumour-Associated and Differentiation Antigens Associated with Type 2 Backbone Sequences

In the sera of patients with the autoimmune haemolytic disorder known as cold agglutinin disease, there occur high titre monoclonal autoantibodies known as anti-I and anti-i cold agglutinins. It so happens that the I and i antigens recognised by these natural monoclonal antibodies are developmentally regulated carbohydrate structures [8, 10]. The i antigen which is a linear oligosaccharide consists of repeating type 2 sequences and is a prominent antigen of human foetal erythrocytes. In contrast it is the branched structure, I antigen, which predominates in the erythrocytes of adults (table I).

The I and i sequences serve as backbone structures for the H, A and B antigens [8, 10] (figure 2) which occur not only on erythrocytes but also in various tissues of the body where their expression is developmentally regulated. Of special interest is the mode of expression of the antigens ABH and I in the gastric mucosa and the mucosal glycoproteins (gastric mucins) [4, 11]: in individuals who are "secretors", (approx. 75% of the population) the gastric mucosa and gastric mucins strongly express the H, A or B antigens. In contrast, in "non-secretors", these antigens are lacking in the superficial gastric mucosa and the mucins they secrete. Instead, the I antigen is strongly expressed; in particular, the I antigenic determinant consisting of the branch point sequence,

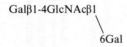

recognised by anti-I Ma (table I, figure 2). On the other hand when "secretors" develop gastric cancer, the I(Ma) determinant is strongly expressed in their tumours; this is presumably due to incomplete biosynthesis of the H, A or B determinants. Thus, the I(Ma) determinant can be regarded as a "distinctive" marker of the gastric mucosa of "non-secretors" and a tumour-associated antigen in the gastric cancer tissues of "secretors" (table I). We are currently measuring the level of the I(Ma) antigen in the gastric juice of "secretors" in order to evaluate its potential as a diagnostic marker of gastric cancer.

The 1–4, 1–6 linked branch can behave as a differentiation antigen in breast-epithelial cells [12] as shown by two hybridoma antibodies M 39 and M 18. These were raised by *Foster et al.*, against human milk fat globules [13]. These two hybridoma antibodies resemble anti-I(Ma) in their specificity for the branch point sequence (table I) and their agglutination reactions with erythrocytes of adults [12]. However, there are differences in the fine specificities of these three antibodies [12, 14] such that they react with only partially overlapping sub-populations of breast epithelial cells.

In the mouse, the I and i antigens behave as early embryonic antigens [15]. The I but not the i antigen is expressed on the earliest mouse embryo; the i antigen appears simultaneously with the onset of differentiation and becomes expressed on the primary endoderm (table I). Thus the antigenic change from i to I during development, as seen with human erythrocytes is not a general phenomenon and the reverse order may occur.

Fucosylated Type 2 Chains May Behave as Differentiation and Tumour-Associated Antigens

The type 2 chain when $\alpha 1-3$ fucosylated as in 3-fucosyl-N-acetyllactosamine, table I, behaves as a differentiation antigen in two systems (table I, figure 2). In the mouse this determinant [16, 17] is the 8-cell stage specific embryonic antigen, known as SSEA-1, recognized by a monoclonal antibody originally raised by *Solter and Knowles* [18] against F9 teratocarcinoma cells of the mouse. However, in man the same trisaccharide structure behaves [19, 20] as a distinctive marker of granulocytes amongst cells of the peripheral blood, as recognized by two monoclonal antibodies, VEP8 and VEP9 raised by *Rumpold et al.* [21]. It appears that approximately 50% of monoclonal antibodies against human myeloid cells recognise this determinant (*H. C. Gooi et al.*, unpublished observations) these include the monoclonal antibody VIM-D5 discussed at this meeting by Dr. Knapp and several other monoclonal antibodies VIM-4, VIM-7, VIM-9, VIM-C6 and VIM-6 raised by *Knapp et al.* [22]. The remaining 50% of the myeloid cell-specific monoclonal antibodies including VIM-2 [22] recognize saccharide structures which are distinct from 3-fucosyl-N-acetyllactosamine (*Uemura et al.,* unpublished observations). A number of other laboratories have also reported monoclonal antibodies against myeloid cells which have specificity for 3-fucosyl-N-acetyllactosamine, as discussed by Dr. Ginsburg at this meeting. Mice immunised with a number of human tumours have also produced antibodies with specificity for this structure [23]. It is of interest that this antigen is not expressed as a granulocyte antigen in the mouse (*S. Thorpe and T. Feizi,* unpublished observations). Thus, apart from individual and tissue differences in the expression of carbohydrate differentiation antigens there are marked species differences.

At later stages of development the Ii and SSEA-1 antigens can no longer be detected in the majority of the tissues of the mouse [15, 18]. In certain epithelial tissues the blood group H antigen is strongly expressed [15]. Since the fucose residues associated with blood group H antigen mask the antigenicity of the I, i and SSEA-1 structures [16, 17], we envisage that the apparent "disappearance" of the latter antigens may be due to their being masked by additional fucosylation or other substitutions. On the other hand the resulting difucosyl type 2 chains (table I, figure 2) express a tumour-associated antigen of human colonic epithelium as recognised by the hybridoma antibody C 14 [24]. This antigen is, in addition, expressed as an intracellular antigen of human granulocytes ([24] and *Brown et al.,* unpublished observations).

The interrelationships of these several antigens including the type 2 based blood group H, A and B antigens are illustrated in figure 2.

The Type 1 Backbone Structure as a Marker of Undifferentiated Teratocarcinomas and Foetal Endodermal Tissues of Man

Undifferentiated teratocarcinomas and endodermal tissues of the human foetus express an antigen recognized by a hybridoma antibody FC 10.2 [25] which was produced by *Williams et al.* following immunisation of mice with formalin fixed human embryonal carcinoma line LICR LON HT 39/7. We have recently shown [26] that the antigenic determinant recognized by this antibody involves the type 1 based sequence shown in table I. This antibody shows little reaction with endodermal tissues of the adult, although the type 1 sequence is well known to occur on glycoproteins [4] and glycolipids [27] of the gastrointestinal tract. This is because the type 1 sequence becomes $\alpha 1-4$ fucosylated and converted to the Le^a antigen or $\alpha 1-2$ fucosylated and converted to the blood group H antigen or alternatively it may be difucosylated and converted into the Le^b antigen shown in figure 3. These various glycosylations result in the masking of the FC 10.2 determinant. Presumably in the human foetus and in the undifferentiated teratocarcinoma cells a high proportion of the backbones are unsubstituted. The carbohydrate specificity of FC 10.2 antibody resembles that of a human Waldenstroem macroglobulin, IgM^{WOO} [28].

For completeness, it should be pointed out that the colon tumourassociated antigen recognised by the monoclonal antibodies 19-9 produced in the laboratory of Koprowsky has been shown by *Magnani et al.,* [29] to involve the $\alpha 2-3$ sialylated[1] form of the Le^a antigen as shown below:

NeuAcα2-3Galβ1-3GlcNAc
    | 1,4
   Fucα

Like the other tumour-associated antigens the 19-9 determinant is not confined to colonic adenocarcinomas but it has been reported to occur as a normal component of the pancreas [23].

Glycoprotein and Glycolipid Carriers of The Tumour-Associated and Differentiation Antigens

These blood group related carbohydrate structures are well known to be carried on glycoproteins and glycolipids. We have recently compared their expression on glycoproteins and glycolipids of erythrocytes, granulocytes, the promyelocytic cell line HL 60, and mouse teratocarcinoma cells [30, 31]. Immunostaining of (a) total cell lysates after polyacrylamide gel electrophoresis and transfer on to ni-

[1] NeuAc = N-acetylneuraminic acid

trocellulose ("Western" blotting) and (b) glycolipid extracts on thin layer chromatography plates, has revealed that the proportion of glycoproteins and glycolipids carrying these antigens vary in different cell types. Precise quantitations are not possible by the techniques used, however, it can be stated that the I and i antigens are readily detected on glycoproteins as well as glycolipid extracts of erythrocytes and SSEA-1 on those of granulocytes and HL 60 cells. However, in teratocarcinoma cells of the mouse these antigens are detected predominantly on glycoproteins.

Conclusions

Our observations with monoclonal antibodies as well as studies from a number of other laboratories are reinforcing the concept that saccharides of the cell surface are important tumour-associated and differentiation antigens. However, none of the antigens thus far characterized have been uniquely specific for a given cell type or tumour. The changes we are observing seem to represent changes in the absolute amounts and the relative proportions of various carbohydrate structures whose changing expression in growing and differentiating cells is precisely programmed and in normal mature cells is genetically predictable. By inference the changes in the carbohydrate structures that we are observing in the neoplastic cells, represent changes in the activities of glycosyltransferases or of the genes that code for the biosynthesis of these enzymes. The relationship of these changes to the neoplastic process is unclear and requires investigation.

These changes would be extremely difficult to detect without the aid of monoclonal antibodies. While the majority of these monoclonal antibodies may be inappropriate as diagnostic or immunotherapeutic reagents [32], they are of unquestioned value as precise biochemical and immunochemical tools. Through the use of well characterized monoclonal antibodies it can be anticipated that important advances will be made in the biochemistry of the neoplastic and differentiation processes.

Summary

Studies with monoclonal antibodies have revealed that a number of antigens with changing expression during embryogenesis, stages of cell differentiation and oncogenesis are carbohydrate chains of cell surface glycoproteins and glycolipids. However, none of the antigens thus far identified are uniquely specific for a normal or a tumour cell type. One antigen may be tumour-associated in one cell type and a normal antigen in another.

While it can be envisaged that these monoclonal antibodies may be exploited for the detection of tumours in specific organs, clearly their overall value in tumour diagnosis and therapy will be limited. On the other hand, the well characterized monoclonal antibodies will in turn be important biochemical tools for the analysis of complex carbohydrate structures of normal and neoplastic cells.

References

1 Köhler, G; Milstein, C.: Continuous cultures of fused cells secreting antibody of predefined specificity. Nature, Lond. *256:*495–497 (1975).
2 Feizi, T.: Carbohydrate differentiation antigens. Trends Biochem. Sci. *6:*333–335 (1981).
3 Feizi, T.: The antigens Ii, SSEA-1 and ABH are an interrelated system of carbohydrate differentiation antigens expressed on glycosphingolipids and glycoproteins; in Makita, Handa, Taketomi, Nagai (eds.), New vistas in glycolipid research, Advances in experimental medicine and biology, vol. 152, pp. 167–177 (Plenum Publishing, New York, London 1982).
4 Hounsell, E. F.; Feizi, T.: Gastrointestinal mucins. Structures and antigenicities of their carbohydrate chains in health and disease. Med. Biol. *60:*227–236 (1982).
5 Kornfeld, R.; Kornfeld, S.: Structure of glycoproteins and their oligosaccharide units, in Lennarz (ed.) The biochemistry of glycoproteins and proteoglycans, pp. 1–34. The John Hopkins University School of Medicine. (Plenum Publishing, New York, London 1980).
6 Hakomori, S.: Glycosphingolipids in cellular interaction, differentiation and oncogenesis. Annu. Rev. Biochem. *50:*733–764 (1981).
7 Watkins, W. M.: Biochemistry and genetics of the ABO, Lewis and P blood group systems, in Harris, Hirschhorn (eds.) Advances in human genetics, vol. 10, pp 1–136, 379–385 (Plenum Publishing, New York, London 1980).
8 Feizi, T.: The blood group Ii system: a carbohydrate antigen system defined by naturally monoclonal or oligoclonal autoantibodies of man. Immunol. Commun. *10:*127–156 (1981).
9 Picard, J. K.; Feizi, T.: Peanut lectin and anti-Ii antibodies reveal structural differences among human gastrointestinal glycoproteins. Mol. Immunol. (in press).
10 Hakomori, S.: Blood group ABH and Ii antigens of human erythrocytes: chemis-

try, polymorphism and their developmental change. Semin. Hematol. *18:*39–62 (1981).
11 Feizi, T.: Blood group antigens and gastric cancer. Med. Biol. *60:*7–11, (1982).
12 Gooi, H. C.; Uemura, K.; Edwards, P. A. W.; Foster, C. S.; Pickering, N.; Feizi, T.: Two mouse hybridoma antibodies against human milk fat globules recognise the I(Ma) antigenic determinant: Galβ1-4GlcNAcβ1-6-Carbohyd. Res. (in press).
13 Foster, C. S.; Edwards, P. A. W.; Dinsdale, E. A.; Neville, A. M.: Monoclonal antibodies to the human mammary gland. I. Distribution of determinants in non-neoplastic mammary and extra mammary tissues. Virchows Arch. Abt. A Path. Anat. *394:*279–293 (1982).
14 Uemura, K.; Childs, R. A.; Hanfland, P.; Feizi, T.: A multiplicity of erythrocyte glycolipids of the neolacto series revealed by immuno-thin-layer chromatography with monoclonal anti-I and anti-i antibodies. Biosci. Rep. (in press).
15 Feizi, T.; Kapadia, A.; Gooi, H. C.; Evans, M. J.: Human monoclonal autoantibodies detect changes in expression and polarization of the Ii antigens during cell differentiation in early mouse embryos and teratocarcinomas; in Muramatsu, Gachelin, Moscona, Ikawa (eds.), Teratocarcinoma and embryonic cell interactions, pp. 201–215 (Japan Scientific Societies Press and Academic Press, Tokyo 1982).
16 Gooi, H. C.; Feizi, T.; Kapadia, A.; Knowles, B. B.; Solter, D.; Evans, M. J.: Stage specific embryonic antigen SSEA-1 involves α1–3 fucosylated type 2 blood group chains. Nature *292:*156–158 (1981).
17 Hounsell, E. F.; Gooi, H. C.; Feizi, T.: The monoclonal antibody anti-SSEA-1 discriminates between fucosylated type 1 and type 2 blood group chains. FEBS Lett. *131:*279–282 (1981).
18 Solter, D.; Knowles, B. B.: Monoclonal antibody defining a stage-specific mouse embryonic antigen (SSEA-1). Proc. natn. Acad. Sci. USA *75:*5565–5569 (1978).
19 Feizi, T.: In foetal antigens and cancer; in Everett, Whelan (eds.), Ciba Foundation Symp., vol. 96, pp. 216–221 (Pitman Medical, London 1983).
20 Gooi, H. C.; Thorpe, S. J.; Hounsell, E. F.; Rumpold, H.; Kraft, D.; Förster, O.; Feizi, T.: Marker of peripheral blood granulocytes and monocytes of man recognized by two monoclonal antibodies VEP8 and VEP9 involves the trisaccharide 3-fucosyl-N-acetyllactosamine. Eur. J. Immunol. *13:*306–312 (1983).
21 Rumpold, H.; Obexerand, G.; Kraft, D.: Analysis of human NK cells by monoclonal antibodies against myelomonocytic and lymphocytic antigens; in Herbermann (ed.), NK cells and other natural effector cells, vol. 2, pp. 47–52 (Academic Press, New York, London 1982).
22 Knapp, W.; Bettelheim, P.; Majdic, O.; Liszka, W.; Schmidmeier, W.; Lutz, D.: Diagnostic value of monoclonal antibodies to leukocyte differentiation antigens in lymphoid and non-lymphoid leukemias; in Bernard, Bounsell (eds.), Human leukocyte markers detected by monoclonal antibodies (Springer, Berlin, Heidelberg, New York, in press).
23 Hansson, G. C.; Karlsson, K.-A.; Larson, G.; McKibbin, J. M.; Blaszczyk, M.; Herlyn, M.; Steplewski, Z.; Koprowski, H.: Mouse monoclonal antibodies against human cancer cell lines with specificities for blood group and related antigens. Characterization by antibody binding to glysphingolipids in a chromatogram binding assay. J. Biol. Chem. *258:*4091–4097 (1983).
24 Brown, A.; Feizi, T.; Gooi, H. C.; Embleton, M. J.; Picard, J. K.; Baldwin, R. W.: A monoclonal antibody against human colonic adenoma recognizes difucosylated type 2 blood group chains. Biosci. Rep., *3:*163–170 (1983).
25 Williams, L. K.; Sullivan, A.; Mc Ilhinney, R. A. J.; Neville, A. M.: A monoclonal antibody marker of human primitive endoderm. Int. J. Cancer. *30:* 731–738 (1982).

26 Gooi, H. C.; Williams, L. K.; Uemura, K.; Hounsell, E. F.; McIlhinney, R. A. J.; Feizi, T.: A marker of human primitive endoderm defined by a monoclonal antibody involves type 1 blood group chains. Mol. Immunol. *20:* 607–613 (1983).
27 Karlsson, K-A.; Larson, G.: Molecular characterization of cell surface antigens of fetal tissue. Detailed analysis of glycosphingolipids of meconium of a human 0 Le(a–b+) secretor. J. biol. Chem. *256:* 3512–3524 (1981).
28 Kabat, E. A.; Liao, J.; Shyong, J.; Osserman, E. F.: A monoclonal IgM macroglobulin with specificity for lacto-N-tetraose in a patient with bronchogenic carcinoma. J. Immunol. *128:* 540–544 (1982).
29 Magnani, J. L.; Nilsson, B.; Brockhaus, M.; Zopf, D.; Steplewski, Z.; Koprowski, H.; Ginsburg, V.: A monoclonal antibody-defined antigen associated with gastrointestinal cancer is a ganglioside containing sialylated lacto-N-fucopentaose II. J. biol. Chem. *257:* 14365–14369 (1982).
30 Feizi, T.: Carbohydrate differentiation antigens recognised by monoclonal antibodies. Biochem. Soc. Transactions *11:* 263–265 (1983).
31 Childs, R. A.; Pennington, J.; Uemura, K.; Scudder, P. N.; Goodfellow, P.; Evans, M. J.; Feizi, T.: Biochem. J. (in press).
32 Feizi, T.: Monoclonal antibodies point to carbohydrate structures as tumour-associated antigens. Med. Biol. *61:* 144–146 (1983).

Ten Feizi, M. D. Applied Immunochemistry Research Group, Clinical Research Centre, GB-Harrow HA1 3UJ (UK)

Identification of Messenger RNA Coding for Carcinoembryonic Antigen

W. Zimmermann, J. Thompson, F. Grunert, G.-A. Luckenbach, R. Friedrich, S. von Kleist

Institute of Immunobiology of the University of Freiburg, Freiburg i. Br., FRG

Introduction

Experiments done in the early fifties revealed that mice were able to reject tumours transplanted from syngenic donors if the recipient mice had been injected with killed tumour cells prior to transplantation. Obviously, the tumour cells expressed antigens which were not present in normal cells and therefore were recognized by the recipients immune system as "non-self". These observations led to the search for human tumour specific antigens which were hoped to be of great diagnostic and therapeutic value. For this purpose, animals were injected with human tumour cells or tumour extracts. The antisera of the immunized animals were absorbed with the corresponding normal human tissue. With the aid of these antisera several tumour associated antigens could be identified.

One of the most widely used human tumour markers is the carcinoembryonic antigen or CEA, originally defined and isolated by *Gold and Freedman* in 1965 [1]. It is a highly glycosylated protein with a molecular weight of about 180,000. It is synthesized during normal human development mainly in the foetal gut [2]. In adults the CEA gene is normally not expressed at a significant level except in gastrointestinal carcinomas [2, 3, 4] and some other tumours [5, 6]. Although elevated serum levels of CEA can be observed in patients without neoplastic disease [7, 8] the concentration of CEA in the serum is an important parameter in the surveillance of carcinoma patients especially in the postoperative phase [9, 10]. Recently, the presence of CEA on the surface of tumour cells has been exploited to localize tumours and metastases with the aid of radioactively labelled anti-CEA antibodies [11]. In addition, anti-CEA antibodies are now used for immunotherapy of tumours [12].

Despite the extraordinary relevance of CEA in clinical oncology [13] nothing is known about the function of this oncodevelopmental protein [14], the gene of which is mainly expressed during embryogenesis and neoplastic growth. Interestingly, this temporal pattern of gene activity is shared by some of the cellular oncogenes which were recently discovered and are thought to be involved in tumourigenesis in animals and humans. The normal counterparts of these genes seem to play an important role during embryogenesis and differentiation [15].

The physicochemical properties of CEA which was originally defined only immunologically have been studied in some detail [16]. CEA is a rather heterogenous glycoprotein even when isolated from a single tumour [17, 18]. This heterogeneity has been attributed to variations in the carbohydrate composition [17] and aminoacid sequence [19]. Recent findings (*F. G. Grunert*, unpublished results) suggest that a CEA species exists in normal tissues that differs from the CEA found in tumours. There is also evidence that lung and breast tumours contain tissue-specific CEA species [20]. This rather complex picture is even more obscured by the existence of numerous cross-reacting antigens also present in serum and tumour tissues [9, 21]. This situation complicates the characterization of CEA.

Several questions can now be raised, that are relevent for the clinical use of CEA:

Are there several CEA genes or is the heterogeneity of CEA due to different modifications (e. g. glycosylation) of a single CEA gene product?

What is the expression pattern of the CEA gene(s) during differentiation of normal and neoplastic tissue?

The knowledge of an answer to these questions might help the better understanding of the discovery that a significant number of patients without neoplasia has elevated serum CEA levels, or that even patients with metastasizing tumours of the gastrointestinal tract have normal CEA levels.

To define the different CEA molecules, the aminoacid sequence of which is difficult to obtain because of extensive glycosylation of the protein, and to distinguish CEA clearly from the crossreacting antigens, it is highly desirable to study the CEA(s) on the genomic level. As a first step towards a better understanding of CEA in molecular terms and to study the regulation of expression of the CEA gene we have characterized the mRNA which codes for the CEA precursor.

Materials and Methods

Monoclonal and polyclonal antibodies were produced as described earlier [22, 23]. Isolation of mRNA, translational analysis of mRNA, immunoprecipitation and cyanogen bromide cleavage of the immunoprecipitate were performed as described before [23]. The size of CEA mRNA was determined essentially as described by *Scherer et al.* [24].

Results and Discussion

To be able to characterize and purify the CEA-mRNA for cloning purposes, an assay for CEA-mRNA had to be developed. This is commonly done by in vitro translation of total mRNA isolated from the tissue which expresses the protein of interest and by immunoprecipitation of the protein precursor. In our studies, we used monoclonal and polyclonal antibodies as a tool to identify the CEA precursor protein. The monoclonal antibodies were produced using conventional methods by immunizing mice with a perchloric acid extract of liver metastases of a human colon tumour and subsequent fusion of the spleen cells of these animals with murine myeloma cells [22]. The liver metastases contain CEA as well as cross-reacting antigens. Three groups of monoclonal antibodies were obtained. In figure 1 the analysis by SDS-polycrylamide electrophoresis of the proteins is shown, which were precipitated by different monoclonal and polyclonal antibodies from an iodine 125-labelled tumour extract. One group of monoclonal antibodies – exemplified by the monoclonal antibody 12/2/17 – precipitated only one protein with a molecular weight of about 180,000 (figure 1, lane *h*), which was shown to have the same electrophoretic mobility (figure. 1, lane *a*) and the same peptide fingerprint as highly purified CEA (not shown). Obviously, this group of antibodies recognized a CEA-specific determinant and was therefore of greatest interest for us. Members of the two other groups precipitated either CEA plus crossreacting antigens (figure 1,

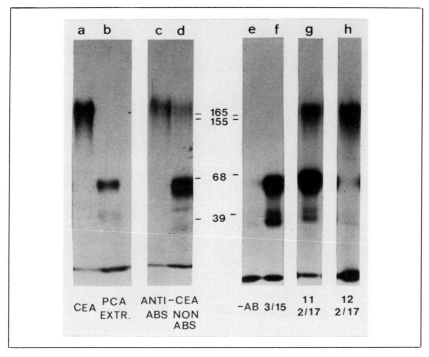

Fig. 1. Characterization of polyclonal and monoclonal anti-CEA antibodies. An iodine-125 labelled perchloric acid (PCA) extract of liver metastases of a human colon tumour was reacted with either monoclonal antibodies 3/15 (lane *f*), 11/2/17 (lane *g*), 12/2/17 (lane *h*) or a polyclonal antibody raised in goat, absorbed (lane *c*), or non-absorbed (lane *d*), with normal lung extract. The immunoprecipitates were analyzed by electrophoresis on a 7.5% sodium dodecylsulfate (SDS) polyacrylamide gel. Lane *a:* highly purified ^{125}J-labelled CEA. Lane *b:* total PCA extract of liver metastases of a human colon tumour. The numbers are the molecular weights of marker proteins in thousands.

lane *g*), or the crossreacting antigens only (figure 1, lane *f*). Also shown are the products precipitated by a polyclonal anti CEA antibody raised in goat before (lane *d*) and after absorption (lane *c*) with an extract prepared from normal lung, which contains crossreacting antigens but essentially no CEA [25].

To identify the CEA precursor, we extracted total RNA from human rectum and colon carcinomas and total tumour mRNA was translated in a rabbit reticulocyte lysate in the presence of [^{3}H] leucine. In figure 2A the electrophoretic pattern of the proteins are

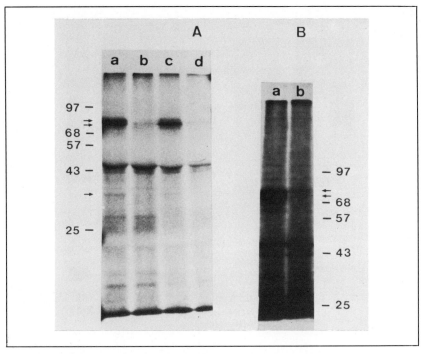

Fig. 2. Electrophoretic analysis of rectum carcinoma mRNA translation products, immunoprecipitated with monoclonal antibody 12/2/17 and a goat anti-CEA antibody. Total tumour RNA of patient 1 (*A*, lanes *a, b; B*, lanes *a, b*) was translated in a rabbit reticulocyte lysate in the presence of tritiated leucine. Immunoprecipitation was performed either with the absorbed polyclonal antibody as in figure 1 (*A*, lanes *a-d; B*, lane *a*) or with the monoclonal antibody 12/2/17 (*B*, lane *b*) in the absence (*A*, lanes *a, c; B*, lanes *a, b*) or the presence (*a*, lanes *b, d*) of an excess of unlabelled CEA. The immunoprecipitates were analyzed by electrophoresis on 10% *(A)* or 7.5% *(B)* SDS-polyacrylamide gels. The arrows point to protein bands, competed for by unlabelled CEA. The numbers are the molecular weights of marker proteins in thousands.

shown which were precipitated by the polyclonal antibody, monospecific for CEA. Addition of an excess of CEA, highly purified from liver metastases of a colon carcinoma, prior to the addition of the anti-CEA antibody prevented the precipitation of two prominent proteins with an apparent molecular weight of about 75,000 and 80,000 (figure 2A, lanes *b, d*). The ratio of these two proteins varied when RNA preparations from different tumours were used (compare in figure 2A lane *a* and *c*). At present we do not know whether the two proteins represent different CEA species coded for by distinct mRNAs, or if their appearance is caused by proteolysis during in vitro translation.

The same two proteins were specifically precipitated – although to a lesser extent – when the monoclonal antibody 12/2/17 was used which recognizes a CEA-specific immunological determinant (figure 2B, lane b).

The capability of monoclonal antibody 12/2/17 to precipitate the CEA precursor demonstrates that it recognizes an epitope on the protein portion of CEA rather than a carbohydrate determinant, because proteins synthesized in a rabbit reticulocyte lysate are not glycosylated [26]. It is supposed that the crossreactivity between CEA and numerous other antigens is caused, at least partly, by shared carbohydrate determinants [27, 28], for example blood group determinants. This assay for CEA mRNA also can be used to improve the specificity of CEA determination by screening for monoclonal antibodies which recognize the protein portion of CEA rather than a probably less specific carbohydrate determinant. We have screened several of our monoclonal antibodies. So far, we only found one that was able to precipitate unglycosylated CEA, which is identical with monoclonal antibody 12/2/17 described above. This however does not imply that all of the other monclonal antibodies react exclusively with carbohydrate determinants, because the threedimensional structure of the protein backbone of the highly glycosylated CEA probably differs from that of the CEA precursor.

CEA isolated from tumours does not contain methionine and, therefore, cannot be cleaved by cyanogen bromide [22, 29], which reacts only with methionine containing polypeptides. To prove the identidy of the putative CEA precursor protein further, we assayed for internal methionine residues. We treated the immunoprecipitate obtained with the polyclonal anti-CEA antibody with cyanogen bromide (figure 3 B). There is no change in size of the putative CEA precursor protein when compared to the control without cyanogen bromide treatment (figure 3 A). In contrast, the unspecifically precipitated protein(s) with a molecular weight of about 57,000 were cleaved (figure 3 A, B).

Taken together, these findings suggest that the proteins with an apparent molecular weight of about 75,000 and 80,000 represent CEA precursor proteins. Taking into account the high carbohydrate content of about 60% of CEA, the observed molecular weight of the CEA

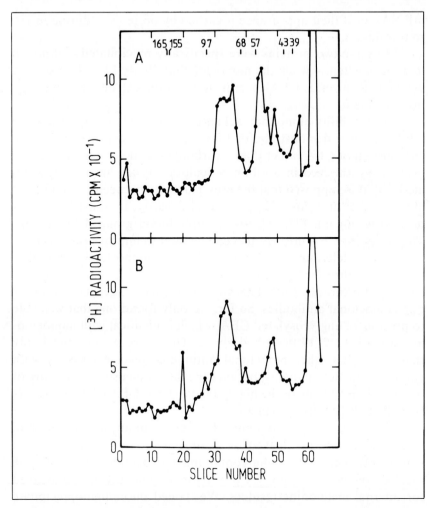

Fig. 3. Electrophoretic analysis of rectum carcinoma mRNA translation products, immunoprecipitated with a goat anti-CEA antibody and reacted with cyanogen bromide. Total RNA isolated from a human rectum carcinoma was translated in a rabbit reticulocyte lysate in the presence of tritiated leucine. Immunoprecipitation with goat anti-CEA IgG was performed. The immunoprecipitate was dissolved in 70% formic acid, divided into two equal portions and incubated in the absence *(A)*, or in the presence *(B)*, of cyanogen bromide. The reaction products were analyzed by electrophoresis on a 7.5% SDS polycrylamide gel. The numbers in *(A)* represent the molecular weights of the marker proteins in thousands.

precursor protein is in good agreement with the reported molecular weight of 180,000 for CEA [30].

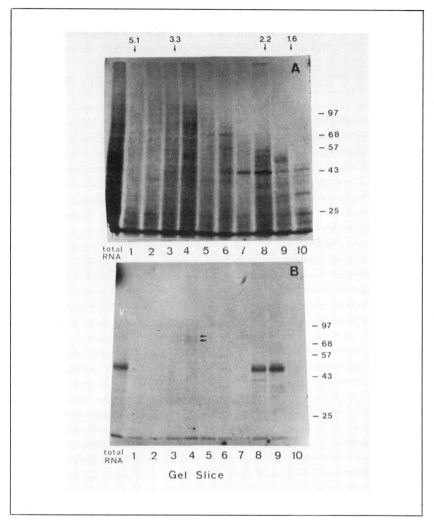

Fig. 4. Size determination of CEA mRNA by electrophoresis on a denaturing agarose gel. Total colon tumour RNA was fractionated by electrophoresis through an agarose gel in the presence of methylmercury hydroxyde. The gel was sliced and the RNA extracted from each slice. The fractionated and total mRNA were translated into proteins and the proteins were analyzed by electrophoresis on SDS-polyacrylamide gels *(A)*. Four fifths of each sample were reacted with the absorbed anti-CEA antibody as in figure 1. The immunoprecipitates were analyzed as above *(B)*. Arrows indicate the position of the CEA precursor proteins. The numbers at the right are the molecular weights of marker proteins in thousands. At the top, position and size of marker DNAs and RNAs in thousand nucleotides are indicated. The dried gels were autoradiographed for one *(A)* and three weeks *(B)*, respectively.

These CEA precursor proteins contain about 0.6% of the total incorporated [^{3}H] leucine. Assuming an average leucine content of the CEA precursor protein and an average translation efficiency of CEA mRNA, this number reflects the relative abundancy of the CEA mRNA. CEA mRNA is, therefore, a middle abundant mRNA species.

To determine the size of CEA mRNA, poly(A)-containing tumour RNA was size-fractionated by electrophoresis through an agarose gel. The gel was sliced, the RNA was extracted from each slice and translated into protein in vitro (figure 4 A). The CEA mRNA-containing fractions were identified by immunoprecipitation with the polyclonal anti-CEA antibody. The largest amount of the CEA precursors was synthesized when RNA from gel fraction four was used (figure 4 B). In this way, the CEA mRNA could be calculated to be about 3,100 nucleotides long. An mRNA of this size can easily code for a protein of a molecular weight of 75-80,000. We have cloned total tumour mRNA and mRNA enriched for CEA mRNA. The CEA-specific cDNA clone will be used as a probe to study the organization and regulation of expression of the CEA gene.

Summary

Total poly(A)-containing RNA extracted from a rectum carcinoma was translated in a rabbit reticulocyte lysate. By addition of either a monoclonal or a polyclonal antibody, both monospecific for carcinoembryonic antigen, two prominent proteins with an apparent molecular weight of about 75,000 and 80,000 were specifically precipitated as shown by electrophoresis of the immunoprecipitate on sodium dodecylsulfate polyacrylamide gels. These proteins behave similar to CEA isolated from liver metastases of a colon tumour in the property of being resistent against cleavage by cyanogen bromide. These findings suggest that these proteins represent CEA precursor proteins. Taking into account the high carbohydrate content of 60% of CEA, the observed molecular weight of the CEA precursor proteins is in good agreement with the reported molecular weight of 180,000 for CEA. By electrophoresis of poly(A)-containing tumour RNA through a denaturing agarose gel, and by in vitro translation fo the fractionated RNA the size of CEA messenger RNA was found to be 3,100 nucleotides.

References

1   Gold, P.; Freedman, S. O.: Demonstration of tumourspecific antigens in human colonic carcinomata by immunological tolerance and absorption techniques. J. exp. Med. *121:*439–462 (1965).

2   Gold, P.; Freedman, S. O.: Specific carcinoembryonic antigens of the human digestive system. J. exp. Med. *122:* 467–481 (1965).
3   Gold, P.; Freedman, S. O.; Shuster, J.: Carcinoembryonic antigen. Historical perspectives, experimental data; in Herberman, McIntire, Immunodiagnosis of cancer, pp. 147–164 (Marcel Dekker, New York, 1979).
4   Kleist, S. von; Burtin, P.: The carcinoembryonic antigen (CEA) and other carcinofetal antigens in gastrointestinal cancers and benign diseases. Prog. Gastroenterol. *III:* 595–615 (1977).
5   Vincent, R. G.; Chu, T. M.; Fergen, T. B.; Ostrander, M.: Carcinoembryonic antigen in 228 patients with carcinoma of the lung. Cancer *36:* 2069–2076 (1975).
6   Steward, A. M.; Zamcheck, D.; Aisenberg, A.: Carcinoembryonic antigen in breast cancer patients: Serum levels and disease progress. Cancer *33:* 1246–1252 (1974).
7   Terry, W. D.; Henkart, P. A.; Coligan, J. E.; Todd, C. W.: Carcinoembryonic antigen: Characterization and clinical applications. Transplant. Rev. *20:* 100–129 (1974).
8   Hansen, H. J.; Snyder, J. J.; Miller, E.; Vandevoorde, J. V.; Miller, O. N.; Hines, L. R.; Burn, J. J.: CEA assay. A laboratory adjunct in the diagnosis and management of cancer. Human Pathol. *5:* 139–147 (1974).
9   Burtin, P.; Gold, P.: Carcinoembryonic antigen. Scand. J. Immunol. *8:* suppl. 8, pp 27–38 (1978).
10 Gold, P.; Shuster, J.: Historical development and potential uses of tumour antigens as markers of human tumour growth. Cancer Res. *40:* 2973–2976 (1981).
11 Mach, J. P.; Carrel, S.; Forni, M.; Richard, J.; Donath, A.; Alberto, P.: Tumour localization of radiolabelled antibodies against carcinoembryonic antigen in patients with carcinoma. New Engl. J. Med. *303:* 5–10 (1980).
12 Goldenberg, D. M.; Gaffar, S. A; Bennett, S. J.; Beach, J. L.: Experimental radioimmunotherapy of a xenografted human colonic tumour (GW-39) producing carcinoembryonic antigen. Cancer Res. *41:* 4354–4360 (1981).
13 National Institutes of Health Consensus Development Conference Statement: Carcinoembryonic antigen: its role as a marker in the management of cancer. Cancer Res. *41:* 2017–2018 (1981).
14 Ibsen, K. H.; Fishman, W. H.: Developmental gene expression in cancer. Biochim. biophys. Acta *560:* 243–280 (1979).
15 Müller, R.; Slamon, D. J.; Tremblay, J. M.; Cline, M. J.; Verma, I. M.: Differential expression of cellular oncogenes during pre- and postnatal development of the mouse. Nature *299:* 640–644 (1982).
16 Pritchard, D. G.; Todd, C. W.: The chemistry of carcinoembryonic antigen; in Herberman, McIntire, Immunodiagnosis of cancer, pp. 165–180 (Marcel Dekker, New York, 1979).
17 Coligan, J. E.; Henkart, P. A.; Todd, C. W.; Terry, W. D.: Heterogeneity of carcinoembryonic antigen. Immunochemistry *10:* 591–599 (1973).
18 Plow, E. F.; Edgington, T. S. Speciation of carcinoembryonic antigen and its immunodiagnostic implications; in Herberman, McIntire, Immunodiagnosis of cancer, pp. 181–239 (Marcel Dekker, New York, 1979).
19 Wang, A. C.; Banjo, C.; Fuks, A.; Shuster, J.; Gold, P.: Heterogeneity of the protein moiety of carcinoembryonic antigens. Immunol. Commun. *5:* 205–210 (1976).
20 Grunert, F.; Luckenbach, G. A.; Haderle, B; Schwarz, K.; Kleist, S. von: Comparison of colon-, lung-, and breast-derived carcinoembryonic antigen and cross reacting antigens by monoclonal antibodies and fingerprint analysis. Ann. N. Y. Acad. Sci. (in press).
21 Kleist, S. von; Burtin, P.: Antigens cross-reacting with CEA; in Herbermann,

McIntire, Immunodiagnosis of cancer, pp. 322–342 (Marcel Dekker, New York, 1979).

22 Grunert, F.; Wank. K.; Luckenbach, G. A.; Kleist, S. von: Monoclonal antibodies against CEA. Comparison of the immunoprecipitates by fingerprint analysis. Oncodevelopment. Biol. Med. *3:* 191–200 (1982).

23 Zimmermann, W., Friedrich, R.; Grunert, F.; Luckenbach, G. A.; Thompson, J.; Kleist, S. von: Characterization of messenger RNA specific for carcinoembryonic antigen. Ann. N. Y. Acad. Sci. (in press).

24 Scherer, G.; Schmid, W.; Atrange, C. M.; Röwekamp, W.; Schütz, G.: Isolation of cDNA clones coding for rat tyrosine aminotransferase. Proc. natn. Acad. Sci. *79:* 7205–7208 (1982).

25 Kleist, S. von; Chavanel, G.; Burtin, P.: Identification of an antigen from normal human tissue that crossreacts with the carcinoembryonic antigen. Proc. natn. Acad. Sci. USA *69:* 2492–2494 (1972).

26 Cowan, N. J.; Milstein, C.: The translation in vitro of mRNA for immunoglobulin heavy chains. Eur. J. Biochem. *36:* 1–7 (1973).

27 Gold, J. M.; Gold, P.: The blood group A-like site of carcinoembryonic antigen. Cancer Res. *33:* 2821–2824 (1973).

28 Holburn, A. M.; Mach, J.-P.; MacDonald, D.; Newlands, M. Studies of the association of the A, B and Lewis blood group antigens with carcinoembryonic antigen (CEA). Immunology *26:* 831–834 (1974).

29 Shively, J. E.; Kessler, M. J.; Todd, C. W.: Amino-terminal sequences of the major tryptic peptides obtained from carcinoembryonic antigen by digestion with trypsin in the presence of Triton X-100. Cancer Res. *38:* 2199–2208 (1978).

30 Slayter, H. S.; Coligan, J. E.: Electron microsocopy and physical characterization of the carcinoembryonic antigen. Biochemistry *14:* 2323–2330 (1975).

Dr. Wolfgang Zimmermann, Institut für Immunbiologie der Universität Freiburg, Stefan-Meier-Str. 8, D-7800 Freiburg i. Br. (FRG)

Antigenic Heterogeneity in Acute Leukemia*

W. Knapp, O. Majdic, P. Bettelheim, K. Liszka, H. Stockinger

Institute of Immunology and 1st Department of Internal Medicine, University of Vienna, Vienna, Austria

Introduction

The immunological characterization of leukemic cells can help us to study the basic principles of normal and malignant leukocyte differentiation and simultaneously contribute to improved diagnostic and therapeutic procedures in clinical medicine. This combination of basic research interests and what seems to be straightforward clinical applicability has attracted the interest of both basic scientists and clinically oriented physicians for many years. For a long time the main obstacle of this approach was the difficulty in raising appropriately specific antisera. This technical problem has been solved by the introduction of the antibody producing hybridoma technology of *Köhler and Milstein* [1].

With this technique it was made possible for the first time to produce monoclonal antibodies (MoAb) of desired specificities and to identify with precision the antigens that they recognize. A new era in the identification and characterization of cells by immunological means was opened.

Human haemopoietic cells were among the first to be intensively studied with this new technology, and the progress so far achieved is impressive.

It is clearly evident, however, that the situation is not as clear and simple as it seems when one reads the current literature. In acute leukemias we frequently observe, for instance, considerable antigenic heterogeneity within individual cell preparations. With the increased number of MoAbs available, the number of immunologically distin-

---

* Supported by Fonds zur Förderung der wissenschaftlichen Forschung in Österreich.

guishable groups and subgroups of acute leukemias is steadily increasing. This heterogeneity of phenotypes makes clinical evaluations difficult. At present we are faced with an increased number of lineage infidelities, i. e. individual leukemic cells which simultaneously express marker characteristics of two different lineages.

These findings, although possibly disturbing in terms of the direct diagnostic and/or therapeutic applicability of monoclonal anti-

Table I. Monoclonal antibodies used in our studies

Designation	Specificity	Reference		
B-cell markers				
VIL-A1	Common ALL antigen	Knapp et al.	1982	[2]
VIB-C5	B-cells and precursors	Bettelheim et al.	(unpublished observations)	
B1	B-cells and precursors	Staskenko et al.	1980	[3]
Y29/55	B-cells	Forster et al.	1982	[4]
Anti IgM	IgM (H-chain specific)	Lansdorp et al.	(unpublished observations)	
T-cell markers				
WT-1	T-cells	Tax et al.	1982	[5]
OKT-1	T-cells	Evans et al.	1981	[6]
9.6	T-cells	Kamoun et al.	1981	[7]
NA1/34	Thymocytes	McMichael et al.	1981	[8]
LEU4	T-Lymphocytes	Ledbetter et al.	1980	[9]
LEU3a	T-Helper	Ledbetter et al.	1980	[9]
LEU2a	T-Suppressor	Ledbetter et al.	1980	[9]
Myeloid markers				
VIM-D5	Myeloid cells, monoblasts	Majdic et al.	1981	[10]
VIM-2	Myelomonocytic cells	Knapp et al.	(in press)	[11]
VIM-D2	Monocytes, monoblasts	Majdic et al.	1982	[12]
MCS-2	Myeloblasts, monocytes	Sagawa et al.	1983	[13]
VIM-12	OKM1 equivalent (GP94/155)	Majdic et al.	(unpublished observations)	
63D3	Monocytes (GP200)	Ugolini et al.	1980	[14]
Erythroid markers				
VIE-G4	Erythropoietic cells Glycophorin A	Liszka et al.	1983	[15]
R18	Erythropoietic cells Glycophorin A	Edwards	1980	[16]
Platelet markers				
AN51	Platelet GPIb	Vainchenker et al.	1982	[17]
J15	Platelet GPIIb/IIIa complex	Vainchenker et al.	1982	[17]
C8-13	Platelet GPIIa	Landsdorp et al.	(unpublished observations)	
C17-27	Platelet GPIIIa	Landsdorp et al.	(unpublished observations)	
"Blast" markers				
VID-1	HLA-DR	Majdic et al.	(unpublished observations)	
VIP-1	Transferrin receptor	Reinherz et al.	1980	[18]
OKT-10	Activated T-cells	Reinherz et al.	1980	[18]

bodies cannot be neglected. We should rather try to learn from these peculiarities and seek solutions. In essence these problems are certainly not unique for acute leukemias and are probably equally true for other poorly differentiatied forms of tumours. Acute leukemias which are much easier to study can therefore give informations which are of general importance in tumour biology.

Materials and Methods

Monoclonal Antibodies

The monoclonal antibodies which are routinely used in our laboratory for the characterization of leukemic cells are listed in table I.

Mononuclear Cells from Leukemia Patients

Mononuclear cells were isolated as described before [10]. The diagnosis of leukemia was made using standard clinical, morphological and cytochemical criteria. All neoplastic preparations selected for this study showed >75% abnormal cells.

Immunofluorescence

The reactivity of monoclonal antibodies with leukemic cells was tested in indirect immunofluorescence using, as secondary reagent, FITC labelled goat $F(ab')_2$ antibodies against mouse IgG + IgM. It was essential for our studies that the conjugate was previously extensively absorbed with human Ig to remove crossreactive antibodies.
All readings were performed with a Leitz Ortholux fluorescence microscope equipped with a Ploem Opak II illuminator (Leitz, Wetzlar, FRG) and in some instances also with the fluorescence activated cell sorter FACS 440 (Becton Dickinson, Sunnyvale, Ca., USA).

Neuraminidase Treatment

Cells (4×10^6/ml) were incubated with 0.2 U/ml Vibrio cholerae neuraminidase (Behringwerke AG, Marburg, FRG) in a shaking water bath at 37 °C for 30 min. The reaction was terminated by washing three times with PBS.

Results

Heterogeneity of Antigen Expression within Individual Cell Samples

The proportion of antibody reactive blast cells in individual cell samples of leukemia patients varies considerably. Particularly outspoken is this variation in AML. It can also be found, however, in ALL.

In figure 1 the percentage distribution of VIL-A1 (= anti CALLA MoAb; [2]) reactive cells in 47 samples of Non T-Non B cell ALL patients is shown. The majority of samples are in the range between 80 and 100% or are completely negative. There are, however, also quite a few samples with percentage values between 30 and 70%. These values cannot solely be explained by an admixture of non-leukemic cells, since samples with less than 90% blasts were excluded from this analysis.

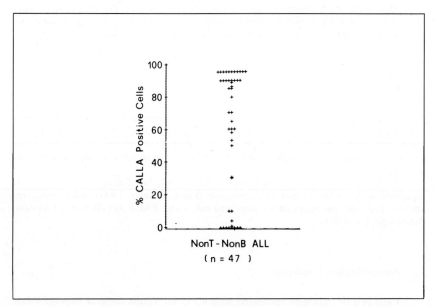

Fig. 1. Percentage distribution of VIL-A1 (= anti CALLA) reactive cells in NonT-NonB ALL samples.

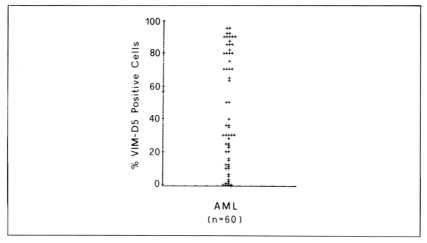

Fig. 2. Percentage distribution of VIM-D5 (= anti myeloid) reactive cells in AML.

The percentage distribution of reactive cells was even more heterogeneous in AML. This was not only true for the myeloid antigen defined by the VIM-D5 antibody [10], but also for HLA-DR antigen expression (figures 2 and 3, respectively).

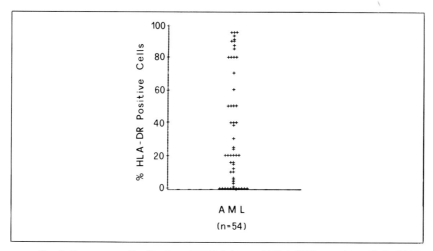

Fig. 3. Percentage distribution of VID-1 (= anti HLA-DR) reactive cells in AML.

Table II. Major acute lymphoblastic leukemia (ALL) subgroups as defined by monoclonal antibodies

Marker profile								Designation[a] of subgroups	Approximate incidence in ALL %
TdT	VILA1	VIBC5	Cμ	S-IgM	9.6	OKT6	OKT3		
+	−	−	−	−	−	−	−	UALL	12
+	+	+	−	−	−	−	−	CALL (pre pre B)	52
+	+	+	+	−	−	−	−	CALL (pre B)	13
−	−	+	−	+	−	−	−	B-ALL	3
+	−	−	−	−	+	−	−	T-ALL (Prothymoc.)	16
+	−	−	−	−	+	+	−	T-ALL (Cortic. Thym.)	4

[a] UALL = undifferentiated ALL, CALL = common ALL, B-ALL = ALL of B cell type, T-ALL = ALL of T cell type.

Table III. Antigenic profiles observed in acute lymphoblastic leukemias of T-cell type (note the heterogeneity)

Pat.	WT-1	OKT-1	9.6	NA1/34	LEU4	LEU3a	LEU2a	TdT	VIL-A1
K. E.	90[a]	80	10	0	0	0	0	pos.	0
W. G.	70	80	0	0	0	0	0	pos.	0
H. E.	90	n. t.	60	0	n. t.	n. t.	n. t.	neg.	0
H. H.	n. t.	95	95	0	0	0	0	pos.	0
S. R.	40	80	70	10	0	0	0	pos.	0
B. R.	25	40	90	20	0	0	0	pos.	0
H. G.	80	80	80	80	10	50	0	pos.	0
I. A.	n. t.	n. t.	0	60	0	0	55	pos.	0
B. A.	50	n. t.	80	50	n. t.	n. t.	n. t.	pos.	0
K. D.	n. t.	n. t.	25	40	35	90	70	pos.	0
S. H.	80	90	90	60	90	75	35	pos.	0
L. H.	90	n. t.	90	0	0	80	75	pos.	0
E. E.	80	80	50	0	0	70	80	pos.	0
S. K.	50	80	40	0	80	0	60	pos.	0
S. R.	n. t.	90	0	0	90	0	0	pos.	0
L. M.	90	70	0	0	0	0	0	pos.	80
F. R.	n. t.	n. t.	95	15	0	0	0	pos.	40
K. E.	90	90	80	n. t.	0	0	0	n. t.	80
K. O.	n. t.	75	85	90	0	90	90	pos.	26
E. S.	n. t.	n. t.	90	40	20	10	n. t.	pos.	60

[a] Percent reactive cells.

Phenotype Heterogeneity

In parallel with the increased number of MoAbs available, the number of immunologically distinguishable groups and subgroups of acute leukemias is steadily increasing. With the MoAbs used in our laboratory for the typing of acute leukemias (listed in table I) we can, for instance, distinguish six major subgroups of ALL (see table II), which can on the basis of the reaction pattern of the leukemic cells with the various subset specific antibodies still further be subdivided into even smaller entities of seemingly identical cell populations. This phenomenon is most outstanding in acute leukemias of T cell type (see table III). It is here that the most outspoken heterogeneity seems to exist, in this respect.

Biphenotypes

Increasingly we also observe socalled biphenotypic leukemias more frequently, i. e. acute leukemias with immunological markers of two or more different hemopoietic lineages. In some instances these markers are coexpressed by the same individual cells. Other cases represent typical double leukemias, in which two distinct blast cell types with typical immunological and sometimes also morphological features are present. A summary of our findings in this respect is shown in table IV.

Sialylation-Dependent Variations

In several instances sialic acid residues are involved in the determinants recognized by differentiation and/or lineage specific MoAb. This may lead to completely controversial situations.

If sialic acid residues are a part of the antigenic determinant, the reactivity of the respective antibody is dependent upon the degree of

Table IV. Distribution of lymphoid and myeloid markers in 298 acute leukemias

		Lymphoid markers	
		+	−
Myeloid	+	29	140
Markers	−	113	16

sialylation of the respective glycoprotein or glycolipid. This behaviour can influence the cellular reaction pattern of a given antibody.

The reaction patterns of two anti glycophorin-A antibodies with three haemopoietic cell lines with erythroid features (K-562, K-562S and HEL) can serve as such an example. Both antibodies specifically recognize the erythroid specific major cell surface glycoprotein glycophorin A. One of them, VIE-G4, is directed against a neuraminidase sensitive determinant of glycophorin A [15]. The other one, R18, recognizes a protein determinant [16].

When recently tested in parallel with the above mentioned cell lines it turned out, that VIE-G4 reacts only with 10% of K-562/4 cells and 28% of HEL cells. However, it recognized 60% of K-562S cells. The proportions reactive with R18 were 64% for K-562/4, 50% for HEL and 63% for K-562S.

In K-562S the glycophorin A positive cells are thus recognized by both antibodies, while in the K-562/4 and HEL cell preparations studied in these experiments only a proportion of the glycophorin A positive cells reacted also with VIE-G4. This indicates, that the glycophorin A molecules on K-562S cells are fully glycosylated, whereas this is not the case with the K-562/4 and HEL cells used in these experiments.

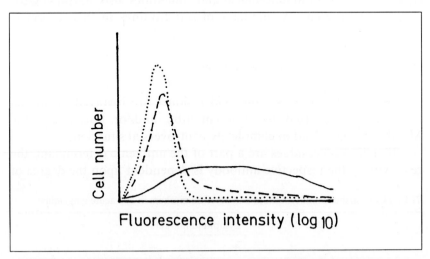

Fig. 4. FACS analysis of untreated (———) as compared to neuraminidase treated (- - - - -) AML cells after staining with the MoAb VIM-D5. .... = negative control.

The other extreme is a situation where a given structure is abundantly present on cell membranes but its reactive sequence is masked in the presence of sialic acid. Such a situation in which sialic acid residues are not involved in the determinant recognized by MoAb but rather interfere with the antibody reactivity is illustrated in figure 4. It shows the FACS analysis of untreated AML cells, as compared to neuraminidase treated AML cells, after staining with a MoAb (VIM-D5) directed against the terminal trisaccharide sequence 3-fucosyl-N-acetyllactosamine (*T. Feizi,* this volume, pp. 51–63). A considerable proportion of previously unreactive cells become clearly and strongly VIM-D5 positive after neuraminidase treatment.

The sialic acids are important constituents of glycoproteins and gangliosides on cell surfaces. The degree of sialylation of a given cell surface protein or glycolipid seems to change, however, during differentiation and possibly depends also on the functional state. Antibodies directed against determinants which are also under the influence of sialic acid residues can, therefore, give highly relevant information about the maturation level or the functional level of a given cell type. However, before one can really appreciate the benefits of this information one has to learn the rules of this game, and how it is influenced or aberantly regulated in malignant processes.

Discussion

The complexity of hematopoiesis and the humoral and cellular influences involved in its regulation make this organ system a challenging field for the study of normal and malignant cellular growth and differentiation. It is, therfore, not surprising that the investigation of membrane markers of haemopoietic cells also has a relatively long tradition.

It has been emphasized on many occasions that the heterogeneity of leukemic cells mirrors the various stages of normal haemopoietic differentiation. This conclusion is mainly based on observations with more or less well differentiated haemopoietic malignancies. It is a common finding that apart from an obvious maturation arrest no major phenotypic abnormalities and particularly no major qualitative phenotypic changes are detectable.

If this conservation or fidelity of a qualitatively normal phenotype is also maintained in malignant proliferations of highly immature haemopoietic cells, including non-lymphoid precursor cells it can give us an insight into the early stages of haemopoietic development.

One requirement for the study of normal haemopoietic differentiation is the ability to recognize the individual stages of development preceding the mature form of a cell type.

In the more mature stages of development, the recognition of different stages is relatively easy on the basis of cell morphology, marker characteristics and function.

The recognition of cells in the early stages of development back to the stem cell is increasingly difficult because of the lack of easily recognizable characteristic features of these cells. In addition, since the development of a cell lineage occurs via a process of exponential growth, the number of early progenitor cells needed for homeostasis is small.

Accumulations of immature precursor cells as found in acute leukemias are, therefore, ideal models and every attempt should be made to systematically map the informative nuclear, cytoplasmic and surface structures of these cells. On the basis of these data it will also be possible to test for acute leukemias the hypothesis that any leukemic cell has a normal counterpart. If affirmative the results will allow the definition and dissection of the early stages of differentiation.

We, therefore, began to systematically evaluate the immunological phenotype of acute leukemia cells with a battery of monoclonal antibodies. In this paper we would like to discuss some of our findings and in particular the observed antigenic heterogeneities in acute leukemias.

The first aspect we were confronted with, was the heterogeneity of antigen expression within individual cell samples from ALL and AML patients. Three such examples are shown in figures 1–3. If one considers the clonal nature of leukemic cells one should not expect such a heterogeneity unless the expression of these antigens is cell cycle or differentiation stage dependent. So far, there is no evidence of a cell cycle dependence of the antigens under study, but it is evident that all the antigens studied by us are differentiation antigens. The most likely explanation for our finding is, therefore, that within individual cell samples from leukemia patients a considerable heterogeneity of differentiation stages does exist, and that the frequently cited maturational arrest is not as precisely located and accurately maintained as frequently postulated. This hypothesis is testable and should be tested.

The same explanation can of course also be offered for the observed heterogeneity of phenotypes in acute leukemia which is responsible for the ever increasing number of immunologically distinguishable leukemia subgroups. However, if gradual differences in the differentiation stages are responsible for these phenotypic variants, one should expect to find a normal counterpart cell for each of these phenotypes.

This is certainly not the case for all phenotypes that we have so far observed.

Particularly in acute leukemias of the T cell group (see table III) antigenic surface patterns cannot infrequently be encountered which have so far not been observed on normal counterpart cells. We are not aware of normal T cells for instance which express the Leu4 and OKT1 antigen but lack the E-receptor (reactivity with 9.6) as well as Leu2a and Leu3a, just to mention one of the examples in table III.

It may well be that such a cell does exist in normals and the study of malignomas can help us a lot to find such particular cell types. It could also be, however, that abnormal gene expressions are more frequent in acute leukemias than generally accepted.

The latter assumption would also be supported by the not too infrequently observed lineage infidelities in acute leukemia samples (see table IV). So far, we have never observed doubly marked cells in normal or regenerating bone marrow. We, therefore, believe that simultaneous expression of myeloid and lymphoid markers by leukemic cells is not a normal phenomenon.

We have to envisage, however, that the great advantage of monoclonal antibodies, their absolute specificity for a given epitope, is at the same time also a danger. Since already the slightest variations of the epitope itself, or of the immediate neighbourhood, may be sufficient to completely abolish or block the determinant, and thus the reactivity of the MoAb, we must be very careful when identifying cells solely on the basis of MoAb reactivity or non-reactivity. How delicate the regulation of antigen expression in some instances can be is shown in this paper with the example of a neuraminidase sensitive determinant and a determinant in which sialic acid residues are not involved in the epitope but rather interfere with MoAb reactivity. The degree of sialylation can therefore markedly influence the expression of these two antigens. Similar situations may well also exist for other antigenic systems and may lead to completely unexpected findings with neoplastic cells.

Acknowledgments: We wish to thank Mrs. *Brigitte Fischer-Colbrie,* Mrs. *Tina Boeckholdt,* and Mrs. *Susanne Beranek* for their skillful technical assistance.

References

1 Köhler, G.; Milstein, C.: Continuous cultures of fused cells secreting antibody of predefined specificty. Nature 256: 495–497 (1975).
2 Knapp, W.; Majdic, O.; Bettelheim, P.; Liszka, K.: VIL-A1, a monoclonal antibody reactive with acute lymphatic leukemia cells. Leuk. Res. 6: 137–147 (1982).
3 Stashenko, P.; Nadler, L. M.; Hardy, R.; Schlossman, S. F.: Characterization of a human B lymphocyte-specific antigen. J. Immunol. 125: 1678–1685 (1980).
4 Forster, H. K.; Gudat, F. G.; Girard, M. F.; Albrecht, R.; Schmidt, J.; Ludwig, C.; Obrecht, J-P.: Monoclonal antibody against a membrane antigen characterizing leukemic human B-lymphocytes. Cancer Res. 42: 1927–1934 (1982).
5 Tax, W. J. M.; Willems, H. W.; Kibbelaar, M. D. A.; De Groot, J.; Capel, P. J. A.; De Waal, R. M. W.; Reekers, P.; Koene, R. A. P.: Monoclonal antibodies against human thymocytes and T lymphocytes; in Peeters (ed.), 29th colloquium 1981. Protides of the biological fluids, pp. 701–704 (Pergamon Press, Oxford 1982).
6 Evans, R. L.; Wall, D. W.; Platsoucas, Ch. D.; Siegal, F. P.; Fikrig, S. M.; Testa, C. M.; Good, R. A.: Thymus-dependent membrane antigens in man: inhibition of cell mediated lympholysis by monoclonal antibodies to T_{H2} antigen. Proc. natn. Acad. Sci. USA 78: 544–548 (1981).
7 Kamoun, M.; Martin, P. J.; Hansen, J. A.; Brown, M. A.; Siadak, A. W.; Nowinski, R. C.: Identification of a human T lymphocyte surface protein associated with the E-rosette receptor. J. exp. Med. 153: 207–212 (1981).
8 McMichael, A. J.; Pilch, J. R.; Galfre, G.; Mason, D. Y.; Fabre, J. W.; Milstein, C.: A human thymocyte antigen defined by a hybrid myeloma monoclonal antibody. Eur. J. Immunol. 9: 205–210 (1979).
9 Ledbetter, J. A.; Evans, R. L.; Lipinski, M.; Cunningham-Rundles, C.; Good, R. A.; Herzenberg, L. A.: Evolutionary conservation of surface molecules that distinguish T lymphocyte helper/inducer and T cytotoxic/suppressor subpopulations in mouse and man. J. exp. Med. 153: 310 (1980).
10 Majdic, O.; Liszka, K.; Lutz, D.; Knapp, W.: Myeloid differentiation antigen defined by a monoclonal antibody. Blood 58: 1127–1133 (1981).
11 Knapp, W.; Bettelheim, P.; Majdic, O.; Liszka, K.; Schmidmeier, W.; Lutz, D.: Diagnostic value of monoclonal antibodies to leukocyte differentiation antigens in lymphoid and non-lymphoid leukemias; in Bernard, Boumsell (eds.), Human leukocyte markers detected by monoclonal antibodies (Springer, Berlin, Heidelberg, New York, in press).

12 Majdic, O.; Bettelheim, P.; Liszka, K.; Lutz, D.: Leukämiediagnostik mit monoklonalen Antikörpern. Wien. klin. Wschr. 94: 387–397 (1982).
13 Sagawa, M.; Tatsumi, E.; Sugimoto, T.; Minato, K.; Lok, M. S.; Minowada, J.: Murine monoclonal antibodies (MCS-1 and MCS-2) reactive with human myeloid leukemia cells. Blood (in press).
14 Ugolini, V.; Nunez, G.; Smith, R. G.; Stastny, P.; Capra, J. D.: Initial characterization of monoclonal antibodies against human monocytes. Proc. natn. Acad. Sci. USA 77: 6764–6767 (1980).
15 Liszka, K.; Majdic, O.; Bettelheim, P.; Knapp, W.: Glycophorin A expression in malignant hematopoiesis. Am. J. Hematol. (in press, 1983).
16 Edwards, P. A. W.: Monoclonal antibodies that bind to the human erythrocyte-membrane glycoproteins glycophorin A and Band 3. Biochem. Soc. Trans. 8: 334 (1980).
17 Vainchenker, W.; Deschamps, J. F.; Bastin, J. M.; Guichard, J.; Titeux, M.; Breton-Gorius, J.; McMichael, A. J.: Two monoclonal antiplatelet antibodies as markers of human megakaryocyte maturation: immunofluorescent staining and platelet peroxidasedetection in megakaryocyte colonies and in vivo cells from normal and leukemic patients. Blood 59: 514–521 (1982).
18 Reinherz, E. L.; Kung, P. C.; Goldstein, G.; Levey, R. H.; Schlossman, S. F.: Discrete stages of human intrathymic differentiation: analysis of normal thymocytes and leukemic lymphoblasts of T-cell lineage. Proc. natn. Acad. Sci. USA 77: 1588–1592 (1980).

Prof. Dr. Walter Knapp, Institute of Immunology, University of Vienna, Borschkegasse 8 A, A-1090 Vienna (Austria)

Hodgkin's Disease and So-Called Malignant Histiocytosis:

Neoplasms of a New Cell Type?

H. Stein[1], J. Gerdes[2], H. Lemke[3], H. Burrichter[4], V. Diehl[4], K. Gatter[1], D. Y. Mason[1]

[1] Nuffield Department of Pathology, John Radcliffe Hospital, Oxford, UK
[2] Institute of Pathology, Christian Albrecht University, Kiel, FRG
[3] Institute of Biochemistry, Christian Albrecht University, Kiel, FRG
[4] Department of Haematology and Oncology, Medizinische Hochschule, Hannover, FRG

Introduction

Despite intensive research, Hodgkin's disease remains a mysterious disease. Three principal questions need to be answered:
(1) What is the origin of Hodgkin and Sternberg-Reed cells?
(2) Is Hodgkin's disease a single entity, or rather a syndrome representing several neoplastic disorders arising from different cell types?
(3) What is the role of the numerous different non-malignant cells, e.g. eosinophils, T cells, etc., which are present in high numbers in Hodgkin's disease tissue?

By applying optimised immunoenzyme histochemical methods, in conjunction with a large panel of monoclonal antibodies, we have been able to come closer to an answer to these three questions.

Materials and Methods

Tissue Samples

The source and processing of the tissue samples is described in detail elsewhere [25].

Antibodies and Anti-Immunoglobulin Enzyme Conjugates

Their sources and reactivities are described in table I.

Immunoenzyme Staining

The two- and three-stage immunoperoxidase method and the mouse-alkaline phosphatase anti-alkaline phosphatase (APAAP) technique have been described in detail in several other articles [4, 21, 25].

Enzyme Cytochemical Staining

Peroxidase staining was performed on unfixed sections or cytospins using the method of *Graham and Karnovsky* [10]. Chloroacetate esterase was demonstrated with the method of *Leder* [12].

Results and Discussion

The Origin and Nature of Hodgkin and Sternberg-Reed Cells?

A comparison (table II) of the antigen and enzyme profile of Hodgkin and Sternberg-Reed cells with that of other cell types of the haematolymphoid system reveals that Hodgkin and Sternberg-Reed cells are different from all known cell populations. We concluded from this finding that Hodgkin and Sternberg-Reed cells represent a new, as yet unidentified, cell population, or alternatively an as yet unrecognized differentiation stage of a known cell type [22, 25].

In order to throw light on this question we attempted to produce antibodies specific for Hodgkin and Sternberg-Reed cells by immunizing rabbits with the Hodgkin's disease-derived cell line L428 (fig. 2, established by *Diehl et al.* [5, 6] and *Schaadt et al.* [17]), which is identical in its antigen profile and enzyme content to Hodgkin and Sternberg-Reed cells as analyzed in tissue sections (see table II last column). This led to the production of an antiserum which could be rendered specific for Hodgkin and Sternberg-Reed cells by repeated absorption with unrelated cells [23].

Table I. Polyclonal and monoclonal antibodies used in the present study

Antibody	Specificity	Molecular weight	Reference	Equivalent or identical antibodies
Tu35	HLA-DR	28,000/34,000	[28]	L243 Becton Dickinson (B–D) Dako-HLA-DR
Tol5	All B cells	150,000	[24]	Dako-pan-B
Anti-IgM	IgM	900,000		Dako-IgM, Bethesda Research Laboratory (BRL)
Anti-IgD	IgD		[9]	Dako-IgD, BRL
C3RTo5	C3b receptor	205,000	[8, 26]	Dako-C3b receptor
Anti-Leu-1	All T cells, B-CLL, centrocytic lymphoma and follicular mantle lymphocytes weakly	65–69,000		Dako-T1, OKT1 (ORTHO)
T11/Lyt3	Sheep erythrocyte receptor	55,000	[27, 11]	Anti-Leu-5 B–D
UCHT1	All T cells	19,000	[2]	OKT3 (ORTHO), Leu-4
Anti-Leu-7	Natural killer cells		[1]	
T-ALL 2	Interdigitating reticulum cells, cortical thymocytes		Bethesda Research Laboratory	
NAl/34	Interdigitating reticulum cells, cortical thymocytes, Langerhans cells		[14]	OKT6 (ORTHO)
Anti-monocyte 2	Monocytes, macrophages		BRL	
OKM1	Granulocytes, monocytes, macrophages, natural killer cells		[3]	
S-HCL 3	Macrophages, hairy cell leukemia cells, granulocytes (weakly)	90,000/150–160,000	[20]	
R4/23	Follicular dentritic reticulum cells, splenic marginal zone cells (weakly)		[15]	Dako-DRC1
Anti-lysozyme	Lysozyme		Dakopatts	
Anti-α_1-antitrypsin	α_1-antitrypsin	110,000	Dakopatts	
Ki-1	See text		[19]	
Ki-24	See text		[22]	
Ki-27	See text		[22]	
3C4	Cells of granulopoietic origin and Sternberg-Reed cells		[18, 25]	

Table II. Comparison of the antigenic and enzymatic profile of the most important cell types of the lymphoid tissue with that of Hodgkin (H) and Sternberg-Reed (SR) cells and the Hodgkin's disease derived cell line L428 as detected by immunoperoxidase staining of frozen and/or paraffin tissue sections or cytocentrifuge slides

Markers used	H and SR cells	B cells	T cells	Monocytes/macrophages	IRC[a]	DRC[b]	Granulopoietic cells	Cell line L428
Surface Ig	−	+	−	−[c]	−	+	−	−
To15 (pan B cell)	−	+	−	−	−	−	−	−
C3RTo5 (C3bR)	−	+	−	+	−	+	+	−
OKT11/Lyt3	−	−	+	−	−	−	−	−
UCHT1/T3	−	−	+	−	−	−	−	−
OKM1	−	−	−	+	−	−	(+)	−
Anti-monocyte 2	−	−	−	+	−	(+)	+	−
S-HCL 3	−	−[d]	−[d]	+	−	+	+	−
Lysozyme	−	−	−	+	−	−	+	−
α1-Antitrypsin	−/+	−	−	+	+	(+)	−	−
NA1/34/T-ALL2	−	−	−	−[e]	−	−	+	+
3C4	+/−	−	−	−	−	−	−	−
Peroxidase	−	−	−	−/+	−	−	+	−
Chloroacetate esterase	−	−	−	−/+	−	−	+	−

[a] Interdigitating reticulum cells;
[b] dendritic reticulum cells of lymphoid follicles;
[c] sometimes weakly positive;
[d] reacts with a very small percentage of lymphoid cells, probably of B cell type;
[e] stains weakly macrophage derived epithelioid cells and other macrophage subsets.

In order to identify the individual antibodies contributing to this labelling reaction we then attempted to raise monoclonal antibodies against the L428 cell line. Among the large number of hybridomas obtained, three produced antibodies reactive with L428 cells and Hodgkin and Sternberg-Reed cells in tissue sections, but not with normal B cells, T cells, macrophages, follicular dendritic reticulum cells, and interdigitating reticulum cells, (table III; [22]). Two of these monoclonal antibodies (Ki-24 and Ki-27) were not restricted in their reactivity to Hodgkin and Sternberg-Reed cells, and hence recognized antigens different from those detected by the polyclonal anti-L428 serum. Ki-24 reacted with a number of non-Hodgkin lymphomas of clear cut B and T cell type, but interestingly not with any normal lymphoid cells. Ki-27 recognized (in addition to Hodgkin and Sternberg-Reed cells) endothelial cells, smooth muscle cells and a proportion of epithelial cells.

Monoclonal antibody Ki-l, in contrast to the other two reagents, labelled Hodgkin and Sternberg-Reed cells in all cases of Hodgkin's disease investigated (figs. 3 and 4) but gave no reaction with other cell types or tissues when tested by a two-stage immunoperoxidase method against all normal tissues and organs. However, when the sensitiv-

Fig. 1: Hodgkin's disease, nodular sclerosis, Giemsa stained paraffin section.

Fig. 2: Hodgkin's disease derived cell line L428. May-Grünwald-Giemsa stained cytospin.

Fig. 3. Hodgkin's disease, mixed cellularity. Frozen section stained with the monoclonal antibody Ki-1 using the three-stage immunoperoxidase method. Only Hodgkin and Sternberg-Reed cells are labelled.

Fig. 4. Hodgkin's disease, nodular sclerosis. Frozen section stained with the monoclonal antibody Ki-1 using the APAAP immuno-alkaline phosphatase method. The labelled Hodgkin and Sternberg-Reed cells are preferentially located close to the mantle zone *(MZ)* of the preserved B cell follicle seen on the left.

Fig. 5. Normal tonsil. Frozen section stained with the monoclonal antibody Ki-1 using the multi-layer APAAP method. Large lymphoid cells around the B cell follicle *(F)* are labelled.

Fig. 6. Same staining as in fig. 5 at a higher magnification. Note the large size of the Ki-1-positive cells and their location at the border between the mantle zone of the B cell follicle *(MZ)* and the T-zone *(TZ)*.

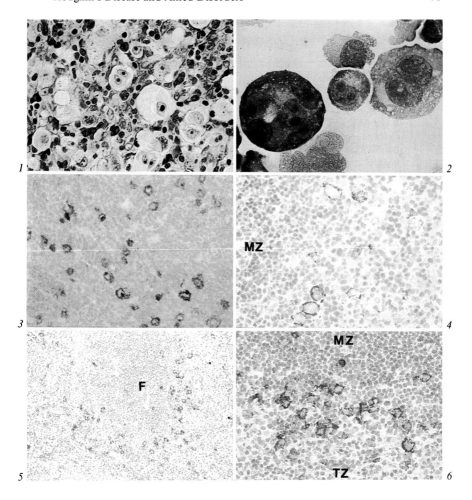

ity of the detection system was enhanced (by application of multi-layer procedures [4, 25]) clusters of large cells around the lymphoid follicles became visible, (the cells of all other biopsies remaining negative) (figs. 5 and 6). These cells are preferentially localized at the border between the B and T cell areas, (although scattered Ki-l-positive cells also occur in the interfollicular T cell rich zone and within lymphoid follicles, mainly at the outer rim of germinal centres). Staining of adjacent tonsil sections with monoclonal antibodies directed against Ig, macrophages, follicular dendritic reticulum cells, interdigitating reticulum cells and natural killer cells revealed that Ki-l-positive cells did

Table III. Reactivity of three monoclonal antibodies that were raised against the Hodgkin's disease-derived cell line L428 and were found to be reactive with the L428 cell line cells, but not with normal B cells, T cells or macrophages

Antibody	Reactive cells (other than H and SR cells)	L428	H and SR cells[a] in tissue sections
Ki-1	Large lymphoid cells around B cell follicles	+ (100%)	+
Ki-24	Cells of various non-Hodgkin's lymphomas, e.g. of centroblastic type, but no normal cells of lymphoid tissue	+ (100%)	+/−
Ki-27	Endothelial cells, smooth muscle cells, epidermal cells, but no normal B or T cells, macrophages, or dendritic or interdigitating reticulum cells or non-Hodgkin's lymphomas with a clear-cut B or T cell phenotype	+ (30–50%)	+/−

[a] H and SR cells – Hodgkin and Sternberg-Reed cells

not react with any of these antibodies. However, a weak staining with antibodies against T cell-associated antigens such as T-11, UCHT1 and antibodies against B cell-associated antigens such as T015 and B1 could not be fully excluded, because of the large numbers of cells which these antibodies stain in normal tissue.

In spite of this uncertainty, the data obtained indicate that normal Ki-1 cells differ from all other known cell types, not only by their expression of Ki-1 and absence of macrophage and accessory cell markers but also in their typical distribution around lymphoid follicles, and their large size. No other cell type described in the literature shows a similar distribution pattern.

Having defined the typical distribution of Ki-1 cells in normal tissue we studied the distribution of Hodgkin and Sternberg-Reed cells (i.e. neoplastic Ki-1 cells) in Hodgkin's disease and found that these cells are also preferentially present around follicles (fig. 4). The similarity of tissue distribution of non-neoplastic and neoplastic Ki-1-positive cells, together with the observation that the perifollicular region is the site in lymphoid tissue where the earliest involvement by Hodgkin's disease is seen [13], strongly suggests that Ki-1 cells occurring in normal lymphoid tissue represent the normal counterpart of Hodgkin and Sternberg-Reed cells.

Table IV. Reactivity of Hodgkin (H) and Sternberg-Reed (SR) cells of Hodgkin's disease of different histological types with the monoclonal antibody Ki-1

Histological type	No. of cases	Cases with Ki-1 positive H and SR cells
Lymphocyte predominance	7	7
Nodular sclerosis	10	10
Mixed cellularity	9	9
Epithelioid cell-rich	3	3
Lymphocyte depletion	2	2
Total	31	31

Is Hodgkin's Disease a Single Entity, or Rather a Heterogenous Syndrome, Representing Several Neoplastic Disorders Arising from Different Cell Types?

As shown in table IV, Hodgkin and Sternberg-Reed cells in Hodgkin's disease of all histological types were reactive with the Ki-1 antibody, suggesting that tumour cells of the same basic nature proliferate in each disease category and that the different types are merely variants of one disorder. If this is correct, the question arises as to the reason for the existence of different histological types of Hodgkin's disease. Since the histological types of Hodgkin's disease differ mainly in the cellular composition of the admixture of non-malignant cells, and in the amount of collagen fibre bundles present, this question is linked to the third question raised.

What is the Role of the Admixture of Various Non-Malignant Cells Present in Large Numbers in Hodgkin's Disease?

It was, and is, widely believed that in Hodgkin's disease the presence of non-malignant cells of various types, usually exceeding the number of tumour cells by a factor of more than 50–200-fold, is the result of a host reaction against the tumour cells, i.e. the Hodgkin and Sternberg-Reed cells. However if this were true, other neoplasms might be expected to show a similar admixture of various cell types, among which the tumour cells would represent only a small minority. Such neoplasms (except Hodgkin's disease) do not, to the best of our knowledge exist, arguing against the concept of a florid host versus tumour reaction as the explanation of the characteristic histological picture seen in Hodgkin's disease. It is more logical to assume that the presence of the non-malignant cells is due to a secretion of soluble factors (i.e. lymphokines) by the Hodgkin and Sternberg-Reed cells. In support of this view recent functional studies have provided evidence that the supernatant of the Hodgkin's disease derived cell line L428 contains lymphokines such as interleukin 1, granulocytic colony stimulating factor [7] and growth factors (*Heit,* personal communication), and other factors, some of which were demonstrable in high concentration.

In this context it is probably of relevance that Hodgkin's disease usually does not begin with a replacement of normal tissue by the tumour cells (as is the case in other malignant tumours), but rather with a severe disorganization of the existing tissue architecture, i.e. a disturbance of cellular composition and arrangement. The heavy infiltration of the involved tissue with non-neoplastic polymorphs, macrophages, plasma cells, T cells and sometimes B cells is well known. Less well known is the fact that the disorganization also involves accessory cells, i.e. the B zone specific follicular dendritic reticulum cells and the T zone specific interdigitating reticulum cells may occur in close proximity (a situation which is never observed in normal lymphoid tissue). Furthermore the ratio of these two cell types may be strikingly altered in favour of one or the other.

Our interpretation of this characteristic pattern of tissue disorganization in Hodgkin's disease is that it does not represent a host-versus-tumour reaction, but rather is induced by the Hodgkin and Sternberg-Reed cells themselves. In this context our observation that the number of Hodgkin and Sternberg-Reed cells in tissue affected by Hodgkin's disease is sometimes lower than the number of Ki-1 cells in normal lymphoid tissue may be of interest. This suggests that, at least initially, the disease symptoms may be caused by a dysfunction of the neoplastic Ki-1 cells rather than by their neoplastic growth. A parallel may be drawn with the syndromes produced by small neoplasms of neuroendocrine origin or (e.g. Zollinger-Ellison syndrome, carcinoid syndrome).

On the basis of this evidence we have drawn four conclusions:
(1) Ki-1 cells are likely to be capable of producing and secreting lymphokines which regulate the cytological composition and the cellular arrangement of lymphoid tissue.

(2) Hodgkin's disease is a neoplasm derived from a Ki-1 cell. These neoplastic Ki-1 cells usually secrete excessive amounts and/or abnormal combinations of lymphokines. It is the secretion of lymphokines, at least early in the process of disease, and less the growth of neoplastic cells per se, which is responsible for the patient's symptoms.

(3) The different histological types of Hodgkin's disease do not represent neoplasms derived from separate cell types, but rather probably represent the secretion of different combinations of lymphokines by the neoplastic Ki-1 cells.

(4) If conclusions (1) and (2) are correct, then neoplasms consisting exclusively of Ki-1 cells should exist, representing a proliferation of Ki-1-positive cells which do not secrete significant amounts of lymphokines.

These four conclusions prompted us to investigate large cell lymphoid tumours lacking surface Ig, B cell, T cell or macrophage-associated antigens for their reactivity with Ki-1. All but one of the 8 large cell lymphomas with such an "antigen silent" immunophenotype studied proved to be Ki-1-positive (fig. 8). Morphologically these cases had initially been classified variously as polymorphic immunoblastic lymphoma, anaplastic carcinoma or, most frequently, malignant histiocytosis (fig. 7). Some of these cases contained foci of multinucleated giant cells morphologically resembling Sternberg-Reed cells. These cells were often positive with the anti-granulocyte monoclonal antibody 3C4, like Sternberg-Reed cells of classical cases of Hodgkin's disease.

These seven Ki-1-positive solid large cell lymphomas appear to represent a pure proliferation of neoplastic cells having an immunophenotype identical to that of classical Hodgkin and Sternberg-Reed cells. This finding can be most easily explained on the basis of the postulate (see above) that the normal cell from which Hodgkin and Sternberg-Reed cells derive can give rise to two morphologically different tumours: (a) to Hodgkin's disease, and (b) to solid large cell lymphomas. These findings also suggested that cases of so-called malignant histiocytosis do not represent neoplasms of the macrophage series, but are tumours closely related to Hodgkin and Sternberg-Reed cells probably derived from the perifollicular Ki-1-positive cell population.

Encouraged by these observations and conclusions we investigated systematically all types of lymphomas for the expression of Ki-1. These studies confirmed that all B cell lymphomas of low grade malignancy (B-CLL, prolymphocytic leukemia, hairy cell leukemia, lymphoplasmacytoid immunocytoma, centroblastic-centrocytic lymphoma, centrocytic lymphoma), lymphoblastic lymphoma/ALL of B and T type and all peripheral T cell lymphomas of small cell type were consistently Ki-1 negative [19, 26]. The results obtained immunohistologically from phenotyping large cell lymphomas are given in table V. They showed, to our surprise, that in large cell lymphomas the Ki-1 antigen is often co-expressed with T cell-associated antigens and/or the B cell-associated antigen To15. Surface Ig was also occasionally found on these Ki-1-positive large cell lymphomas. None of

Table V. Immunohistological labelling reactions of 83 large cell lymphomas

Immunologic categories of large cell lymphomas	No. of cases	Ki-1	α_1-Antitrypsin	Lysozyme	Macrophage associated antigens
HLA-DR+, other antigens absent	8	7	3	0	0
SIg+, To15+, HLA-DR+	35	2	0	0	0
SIg−, To15+, HLA-DR+	15	4	1	0	0
T11/UCHT1+, HLA-DR−	7	0	0	0	0
T11/UCHT1+, HLA-DR+	15	13	4	0	0
To15+, T11/UCHT1+, HLA-DR+	3	3	1	0	0

Table VI. Reactivity of Hodgkin's disease derived cell lines (L428, L540 and L591), EBNA+ lymphoblastoid cell lines (LCL), the Burkitt-lymphoma derived cell line DAUDI and histiocytic cell line U937

	L428	L540	L591	LCL	DAUDI	U937
Ki-1	+	+	+	+	−	−
Ki-24	+	−	−	+	−	−
Ki-27	+	−	−	−	−	−
3C4	+	+	−	+	−	−
OKT11/Lyt 3	−	+	+	−	−	−
UCHT1/T3	−					
HLA-DR	+	+	+	+	+	−
To15	−	+	−	+	+	−
Surface Ig	−	−	−	+	+	−
Antimonocyte 2	−	−	−	−	−	+
S-HCL 3	−	−	−	−	−	+
Lysozyme	−	−	−	−	−	+

these cases expressed lysozyme or other macrophage associated antigens, with the exception of alpha-1-anti-trypsin (see below).

These unexpected findings should be interpreted in the light of results obtained by immuno-phenotyping cell lines (table VI). Two recently developed Hodgkin's disease-derived cell lines L540 and L591 [6] express Ki-1 either in association with T cell antigens alone (L591) or in association with both T cell-associated antigens and B cell-associated antigens (L540) [7]. Among the various other cell lines tested, only EBV-positive lymphoblastoid cell lines proved to be Ki-1-positive. These results, in conjunction with the data obtained in the large cell lymphomas, suggests that·

(1) Ki-1 may be expressed by cells of at least two lineages, namely (a) surface Ig-positive B cells and (b) surface Ig-negative, HLA-DR-positive cells which often express T cell-associated antigens alone or in association with the B cell-associated antigen To15;

(2) there may be cases of Hogdkin's disease in which Hodgkin and Sternberg-Reed cells express, in addition to Ki-1, T cell-associated antigens alone or together with To15.

We therefore re-investigated a larger number of cases of Hodgkin's disease for their reactivity with antibodies against T cell- and B cell-associated antigens and other antigens using optimized immune

---

Fig. 7. Large cell lymphoma that had been diagnosed as malignant histiocytosis. Giemsa stained paraffin section.

Fig. 8. Same case as in figure 7. Frozen section stained with the monoclonal antibody Ki-1 by the APAAP method. The great majority of neoplastic cells are strongly labelled and contrast with unstained residual normal lymphoid tissue (upper left corner).

Fig. 9. Hodgkin's disease with clusters of Hodgkin's cells. Giemsa stained paraffin section.

Fig. 10. Same case as in figure 9. Frozen section stained with the monoclonal antibody OKT11 (anti-SRBC receptor). In addition to the staining of the small T cells there is a weak but distinct staining of the Hodgkin cells. Other antibodies (e.g.: OKT8, anti-IgM, anti-IgD, T015, anti-monocyte 2 etc.) failed to stain the Hodgkin cells.

Fig. 11. Lymph node partly infiltrated by a Ki-1-positive, HLA-DR-positive, OKT11-positive large cell lymphoma. Frozen section stained with the monoclonal antibody Ki-1. Note the "homing" of tumour cells around the B cell follicle *(F)*.

Fig. 12. Lymph node showing a selective infiltration of the marginal sinus by a Ki-1-positive, HLA-DR-positive, OKT11-positive large cell lymphoma.

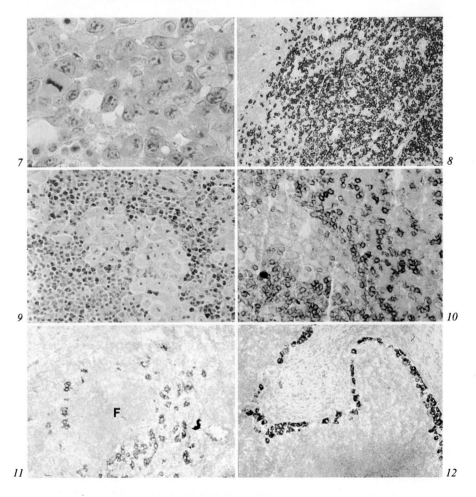

enzyme histological methods. In this context, Hodgkin's disease biopsies containing clusters of Hodgkin cells (fig. 9) were particularly valuable because in sections of the biopsies the surface staining of the Hodgkin's cells could be easily evaluated and was not obscured by the presence of T cells between and around the Hodgkin's cells. The results we obtained revealed four different possible phenotypes of H and SR cells (see table VII). The most common phenotype was the one we have described previously [25], i.e., Ki-1-positive and negative for all other markers except HLA-DR. However, in our first study the

Table VII. Surface antigen pattern on Hodgkin and Sternberg-Reed cells as revealed with immunoenzyme histological methods with increased sensitivity

Antigens	Common phenotype	Variant I	Variant II	Variant III[a]
Surface Ig	−	−	−	−
HLA-DR	+	+	+	+
B cell assoc. antigen To15	−	−	+	+
T11 and/or UCHT1/T3	−	+	−	+

[a] This phenotypical variant was also found in the cells of a polymorphic large cell lymphoma that developed from a case of nodular type of Hodgkin's disease with lymphocyte predominance (nodular paragranuloma).

existence of the three rarer phenotypic varients had been overlooked (see table II, fig. 10).

It is evident that the phenotypic variants of Hodgkin and Sternberg-Reed cells are similar or identical to the phenotypic variants found in Ki-1-positive large cell lymphomas referred to above, except for the two surface Ig-positive cases. This overlap in markers substantiates our concept that the tumour cells of the Ki-1 cell lymphomas which express T cell-associated antigens, with or without the B cell-associated antigen To15, are also identical in nature to Hodgkin and Sternberg-Reed cells. There are several other lines of evidence favouring this view: firstly, a significant number of the Ki-1 cell lymphomas express additional markers that we found to be very characteristic for Hodgkin and Sternberg-Reed cells, i.e., Ki-24, Ki-27, granulocytic antigen 3C4 and alpha-1-anti-trypsin. It has been shown by *Payne et al.* [16] and others, and in our own laboratory [25], that H and SR cells in approximately one third of the cases of Hodgkin's disease contain demonstrable amounts of alpha-1-anti-trypsin in their cytoplasm (alpha-1-anti-trypsin has been shown to be consistently absent from all non-Hodgkin's lymphomas of clear-cut B cell or T cell phenotype [25]; *Pallesen,* personal communication). Alpha-1-anti-trypsin granules were also detectable in approximately one third of the Ki-1-positive large cell lymphomas; there was no relation between the presence of alpha-1-anti-trypsin granules and the expression of the T cell-associated and/or B cell-associated antigens.

Secondly, in both Hodgkin's disease and Ki-1-positive large cell lymphomas the tumour cells showed similar homing properties. This could be especially well demonstrated in partially infiltrated lymph nodes since in such lymph nodes the Ki-1-positive lymphoma cells showed a preferential localization around the B cell follicles as do H and SR cells (fig. 11).

Thirdly, Hodgkin's disease and Ki-1-positive large cell lymphomas exhibit a very similar pattern of spread, in that in both conditions the earliest stage of tumour cell metastases often involves marginal sinus invasion (fig. 12).

Taken together the results of our studies suggests that Ki-1 antigen is present on two cell types: (a) a B cell subset which gives rise to lymphoblastoid cell lines and rarely to large cell lymphomas, and (b) a surface Ig-negative and Ki-1-positive perifollicular lymphoid cell population which we designate "Ki-1 cells". This Ki-1 cell population regularly expresses HLA-DR and may also express T cell-associated antigens and/or B cell-associated antigens. If the perifollicular Ki-1 cell population becomes neoplastic it may give rise to at least two morphological types of neoplasms: Hodgkin's disease and solid large cell lymphomas. It is speculated that neoplastic Ki-1 cells produce Hodgkin's disease if they release lymphokines, whereas the same cells will produce large cell lymphomas if they fail to secrete significant amounts of lymphokines.

Further studies are in progress to substantiate this concept.

Acknowledgments: We thank Miss *Heike Asbahr* and Miss *Kirsten Tiemann* for their skillful technical assistance. This investigation was supported by grants from the Deutsche Krebshilfe and the Leukemia Research Fund.

References

1 Abo, T.; Balch, C. M.: A differentiation antigen of human NK and K cells identified by a monoclonal antibody HNK-1. J. Immunol. *127:*1024–1029 (1981).
2 Beverley, P. C. L.; Callard, R. E.: Distinctive functional characteristics of human "T" lymphocytes defined by E rosetting and a monoclonal anti-T cell antibody. Eur. J. Immunol. *11:*329–334 (1981).
3 Breard, J. M.; Reinherz, E. L.; Kung, P. C.; Goldstein, G.; Schlossman, S. F.: A monoclonal antibody reactive with human peripheral blood monocytes. J. Immunol. *124:*1943 (1980).
4 Cordell, J.; Falini, B.; Erber, W. N.; Ghosh, A. K.; Abdulaziz, Z.; MacDonald, S.; Pulford, K. A. F.; Stein, H.; Mason, D. Y.: Immunoenzymatic labelling of monoclonal antibodies using immune complexes of alkaline phosphatase and monoclonal anti-alkaline phosphatase (APAAP complexes). J. Histochem. Cytochem. (in press).

5   Diehl, V.; Kirchner, H. H.; Burrichter, H.; Stein, H.; Fonatsch, C.; Gerdes, J.; Schaadt, M.; Heit, W.; Uchanska-Ziegler, B.; Ziegler, A.; Heintz, H., Sueno, K.: Characteristics of Hodgkin's disease-derived cell lines. Cancer Treat. Rep. *66:* 615–632 (1982).

6   Diehl, V.; Kirchner, H. H.; Schaadt, M.; Fonatsch, C.; Stein, H.; Gerdes, J.; Boie, C.: Hodgkin's Disease: Establishment and characterization of four in vitro cell lines. J. Cancer Res. Clin. Oncol. *101:* 111–124 (1981).

7   Diehl, V.; Burrichter, H.; Schaadt, M.; Kirchner, H. H.; Fonatsch, C.; Stein, H.; Gerdes, J.; Heit, W.; Ziegler, A.: Hodgkin's disease cell lines: characteristics and biological activities. Haematol. Blood Transfusion *28:* 411–417 (1983).

8   Engleman, E. G.; Levy, R.: Immunologic studies of a human T lymphocyte antigen recognized by a monoclonal antibody. Clin. Res. *28:* 502A (1980).

9   Gerdes, J.; Naiem, M.; Mason, D. Y.; Stein, H.: Human complement (C3b) receptors defined by a mouse monoclonal antibody. Immunology *45;* 645–653 (1982).

10 Graham, R. C. Jr.; Karnovsky, M. J.: The early stages of absorption of injected horseradish peroxidase in the proximal tubules of mouse kidney: ultrastructural cytochemistry by a new technique. J. Histochem. Cytochem. *14:* 291–302 (1966).

11 Kamoun, M.; Martin, P. J.; Hansen, J. A.; Brown, M. B.; Siadek, A. W.; Nowinski, R. C.: Identification of a human T lymphocyte surface protein associated with the E-rosette receptor. J. exp. Med. *153:* 207–212 (1981).

12 Leder, L. D.: Die fermentcytochemische Erkennung normaler und neoplastischer Erythropoesezellen in Schnitt und Ausstrich. Blut *15:* 289–293 (1967).

13 Lukes, R. J.: Criteria for involvement of lymph node, bone marrow, spleen and liver in Hodgkin's disease. Cancer Res. *31:* 1755–1767 (1971).

14 McMichael, A. J.; Pilch, J. R.; Galfre, G.; Mason, D. Y.; Fabre, J. W.; Milstein, C.: A human thymocyte antigen defined by a hybrid myeloma monoclonal antibody. Eur. J. Immunol. *9:* 205–210 (1979).

15 Naiem, M.; Gerdes, J.; Abdulaziz, Z.; Stein, H.; Mason, D. Y.: Production of a monoclonal antibody reactive with human dendritic reticulum cells and its use in the immunohistological analysis of lymphoid tissue. J. clin. Path. *36:* 167 (1983).

16 Payne, S. V.; Newell, D. G.; Jones, D. B.; Wright, D. H.: The macrophage origin of Reed-Sternberg cells. An immunohistologic study. J. clin. Path. *35:* 159–166 (1982).

17 Schaadt, M.; Diehl, V.; Stein, H.; Fonatsch, C.; Kirchner, H.: Two neoplastic cell lines with unique features derived from Hodgkin's disease. Int. J. Cancer *26:* 723–731 (1980).

18 Schienle, H. W.; Stein, H.; Müller-Ruchholtz, W.: Neutrophil granulocytic cellantigen defined by a monoclonal antibody – its distribution within normal haemic and non-haemic tissue. J. clin. Path. *35:* 959–966 (1982).

19 Schwab, U.; Stein, H.; Gerdes, J.; Lemke, H.; Kirchner, H.; Schaadt, M.; Diehl, V.: Production of a monoclonal antibody specific for Hodgkin and Sternberg-Reed cells of Hodgkin's lymphoma and a subset of normal lymphoid cells. Nature *299:* 65–67 (1982).

20 Schwarting, R.; Stein, H.; Wang, C. Y.: The monoclonal antibodies S-HCL-1 and S-HCL3 allow the diagnosis of hairy cell leukaemia. (Submitted for publication).

21 Stein, H.; Bonk, A.; Tolksdorf, G.; Lennert, K.; Rodt, H.; Gerdes, J.: Immunohistologic analysis of the organization of normal lymphoid tissue and non-Hodgkin's lymphomas. J. Histochem. Cytochem. *28:* 746–760 (1980).

22 Stein, H.; Gerdes, J.; Schwab, U.; Lemke, H.; Diehl, V.; Mason, D. Y.; Bartels, H.; Ziegler, A.: Evidence for the detection of the normal counterpart of Hodgkin and Sternberg-Reed cells. Haematol. Oncol. *1:* 21–29 (1983).

23 Stein, H.; Gerdes, J.; Kirchner, H.; Schaadt, M.; Diehl, V.: Hodgkin Sternberg-Reed cell antigen(s) detected by an antiserum to a cell line (L428) derived from Hodgkin's disease. Int. J. Cancer 28: 425–429 (1981).
24 Stein, H.; Gerdes, J.; Mason, D. Y.: The normal and malignant germinal centre. Clin. Haematol. 11: 531–559 (1982).
25 Stein, H.; Gerdes, J.; Schwab, U.; Lemke, H.; Mason, D. Y.; Ziegler, A.; Schienle, W.; Diehl, V.: Identification of Hodgkin and Sternberg-Reed cells as a unique cell type derived from a newly detected small cell population. Int. J. Cancer 30: 445–459 (1982).
26 Stein, H.; Lennert, K.; Feller, A.; Mason, D. Y.: Immunohistological analysis of human lymphoma: Correlation of histological and immunological categories. Adv. Cancer Res. (in press).
27 Verbi, W.; Greaves, M. F.; Schneider, C.; Koubek, K.; Janossy, G.; Stein, H.; Kung, P.; Goldstein, G.: Monoclonal antibodies OKT 11 and OKT 11A have pan-T activity and block sheep erythrocyte "receptors". Eur. J. Immunol. 12: 81–86 (1982).
28 Ziegler, A.; Uchanska-Ziegler, B.; Zeutheu, J.; Wernet, P.: HLA-antigen expression at the single cell level on a K562 and B cell hybrid: an analysis with monoclonal antibodies using bacterial binding assays. Somatic Cell Genet. 8: 775–789 (1982).

Prof. Dr. H. Stein, Nuffield Department of Pathology, John Radcliffe Hospital, Level 4, GB-Oxford OX3 9DU (UK)

Renal Cancer – A Model System for the Study of Human Neoplasia

N. H. Bander

Dept. of Surgery – Urology, Laboratory of Human Cancer Serology, Memorial Sloan-Kettering Cancer Center, New York, N. Y., USA

Introduction

Both logistic and ethical problems have restricted the study of basic cancer biology in the species of primary interest: man. As a result, much of our current knowledge is based on studies in laboratory animals. While this approach has greatly enhanced our understanding of neoplasia in general, its relevance to the human situation cannot always be accepted à priori. A case in point is the concept of tumor-specific antigens. Using transplantation assays in inbred animals with chemically or virally induced cancers, many workers [1, 2, 3] unequivocally demonstrated that these cancer cells expressed specific antigens not present on normal cells. The implications of this concept vis à vis human cancer immunotherapy and immunoprophylaxis are, of course, profound. Unfortunately, innumerable attempts and claims over the past 30 years have failed to firmly establish the existence of tumour-specific antigens in spontaneous human cancers [4].

In recent years, advances in tissue culture techniques have allowed the establishment of a large number of diverse types of human tumour cell lines. The study of such cell lines in vitro offers many advantages: study of cells derived from spontaneous human tumours (as opposed to induced animal tumours); the availability of unlimited numbers of cells; the ability to perform multiple, repeated and varied assays over long time intervals; the ability to study metabolic events in viable cells; and the ability to manipulate cells in vitro in ways not acceptable or not possible in vivo.

The important issue of how closely in vitro cell lines resemble their in vivo counterparts has been looked at in several ways. Complete coverage of this subject is beyond the scope of this discussion. There is evidence, however, that in vitro cell lines bear a striking, albeit imperfect, resemblence to the cells from which they were derived [5]. Therefore, human cell lines provide a useful means to approach the study of human cancer from a somewhat different perspective than animal models. Data gleaned from each source can be collated and then rationally applied to the clinical situation.

While in general, most human cancers can be adapted to in vitro growth only very rarely (less than 1%), melanomas have long been recognized as an exception. There are currently several hundred human melanoma cell lines available in vitro. Perhaps because of this, melanoma has become the subject of study in many laboratories around the world. What is less well appreciated is that human renal cancer can also be established in tissue culture with a relatively high success rate.

Human Renal Cancer In Vitro

Attempts to establish human renal cancer cell lines began in our laboratory in 1974. During the period 1976–1982, 291 specimens of renal cancer were received in our laboratory. Two hundred and twelve (73%) of these were derived from the primary site (kidney) while 79 (27%) were from metastatic lesions. Approximately 50% of the specimens placed in culture will survive one passage. Twenty-six of 212 (12%) nephrectomy specimens and 18 of 79 (23%) metastatic lesions have been established as continuous cell lines; that is, they have survived 6 or more passages and appear capable of indefinite growth. The overall success rate of establishing human renal cancer lines since 1976 has been 15% (43/291). Since 1974 we have established 47 human renal cancer lines. Details of methods utilized have been published previously [6].

The 47 established lines were derived from 44 patients. From 3 patients, 2 different lines were established from separate sites. SK-RC-10 and SK-RC-13 were derived from 2 distinct brain lesions excised at 2 different crainiotomies. SK-RC-26A was established from a metastatic lung lesion while SK-RC-26B was derived from a metastatic lymph node. SK-RC-44 was established from a nephrectomy speci-

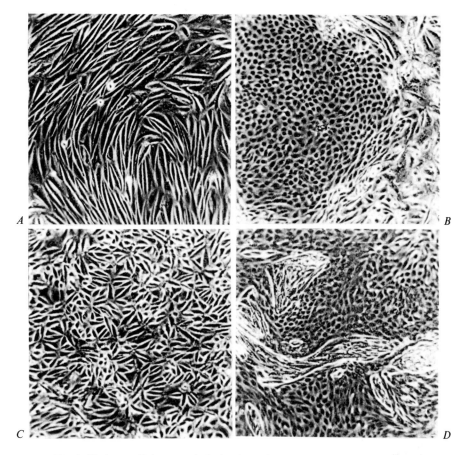

Fig. 1. Various cellular morphologies in a single typical normal kidney culture. *(A)* crescent shaped cells; *(B)* round cells; *(C)* polygonal cells; *(D)* interdigitating growth patterns of two different cell types.

men while SK-RC-45 was established from a metastatic lesion of the contralateral adrenal gland.

Certainly, the ability to grow human renal cancer in vitro is not unique. It is important to emphasize, however, that the success rate of establishing continuous lines of renal cancer is 1–2 orders of magnitude better than other human cancers and equalled only by melanomas. Furthermore, one may also grow normal human kidney epithelium in vitro.

Normal Human Kidney Epithelium In Vitro

Any study of cancer must also examine the normal cellular counterpart in order to provide a point of reference. Normal human kidney epithelium can adapt to short-term culture with a success rate greater than 98%. These lines will grow and divide in vitro for an average of 5 passages and can be maintained for approximately 3 months. If desired, these cells can be cryopreserved and recultured later. The culture medium is identical to that used for renal cancers (Eagle's minimal essential medium supplemented with 1% nonessential amino acids, 2 mM glutamine, 100 U/ml penicillin, 100 µg/ml streptomycin) with the exception that renal cancers grow well in 2.5% fetal calf plus 5.0% newborn calf serum while normal kidney cells require 10% fetal calf serum.

Normal kidney cells in vitro, unlike their neoplastic counterparts, exhibit contact inhibition. In addition, several morphologically heterogeneous cell types can be demonstrated (fig. 1). These appear to reflect the different cell types compromising the nephron (vide infra). Cells demonstrate either crescentic, round or polygonal configurations. Variable growth patterns are also demonstrable with swirls, sheets and nests.

Certainly the most unique feature of the human renal cancer system in vitro is that it is the only tumour system (human or otherwise) in which one can grow and study both the neoplastic cell and its normal cellular counterpart in autologous combinations. The significance of the ability to perform parallel studies on both the viable cancer cell, and its normal counterpart, both derived from the same patient, and to be able to do this on a routine basis simply cannot be over-emphasized. Whether one is studying physiology, genetics, cell surface antigens, oncogenes, etc., this characteristic makes human renal cancer an experimentally highly accessible and useful system.

Murine Monoclonal Antibodies to Human Kidney Antigens

In the Laboratory of Human Cancer Serology at the Memorial Sloan-Kettering Cancer Center, we have produced 36 mouse monoclonal antibodies (mAb) to human kidney antigens. (BALB/c × C57BL/6)F_1 mice are hyperimmunized with viable cells from individual renal cancer lines or normal kidney cultures. Many additional mAb have been raised in parallel studies of melanoma, lung, ovarian, colorectal and breast carcinomas. Clones are selected for analysis by screening supernatants against a panel of viable cell lines using either a mouse mixed hemadsorption assay or, more recently, an enzyme-linked immunosorbent assay (ELISA). Clones producing immunoglobulin which results in an interesting specificity pattern are subloned at least three times at one cell per well. The clones are then expanded by injection subcutaneously to nu/nu mice and/or intraperitoneally into pristane-primed syngeneic mice for the production of sera and ascites respectively. Details of the procedure have been published [7].

Using this high-titered serum or ascites, the in vitro specificity of the antibodies are determined by both direct cell-binding and qualitative absorption assays using an extensive panel of approximately 100 normal and neoplastic cell lines.

Parallel immunochemical studies are performed by metabolically labelling cells with ^{35}S-methionine or ^{3}H-glucosamine. Defined antigens are immunoprecipitated from cell lysates and analyzed by two-dimensional polyacrilimide gel electrophoresis to determine molecular weight and isoelectric point.

Lastly, the sites of antigen expression in vivo are determined by immunofluorescence and immunoperoxidase staining of frozen, nonfixed tissue sections of normal adult, fetal and neoplastic specimens.

The mAb under discussion in this paper, along with the antigens they define, are listed in table I. An abridged in vitro specificity ana-

Table I. Monoclonal antibodies and their defined antigens

mAb (Ig subclass)	Defined Ag	Immunizing cell line	Ref.
S4 (IgG_{2a})	gp160[a]	SK-RC-7 (renal cancer)	[7]
S22 (IgG_1)	gp115	SK-RC-7	[7]
S23 (IgG_1)	gp120r	SK-RC-7	[7]
S27 (IgG_1)	gp120nr	SK-RC-7	[7]
M2 (IgM)	A blood group	SK-RC-28	[7]
S8 (IgM)	B blood group	SK-RC-7	[7]
A99 (IgG_1)	gp170, 140, 140, 28	SK-OV-3 (ovarian cancer)	[8]
C26 (IgG_{2a})	gp40	HT-29 (colon cancer)	[9]
AJ8 (IgG_1)		SK-MG-1 (astrocytoma)	[10]

[a] gp160 = glycoprotein 160 kd

Table II. Abridged monoclonal antibody specificity in vitro as determined by direct cell-binding assays

Cell lines	mAb						
	S4	S22	S23	S27	A99	C26	AJ8
Epithelial cancers							
Renal	25/33	9/33	19/33	31/32	10/10	4/4	–/30
Bladder	–/11	–/11	1/11	3/11	5/5	2/5	–/2
Breast	–/4	–/4	–/4	–/4	2/2	1/1	–/5
Cervix	–/1	–/1	–/1	–/1	1/1		–/1
Colon	–/2	–/2	–/2	–/2	1/1	14/15	–/3
Lung	–/2	–/2	–/1	1/2	2/2	7/8	–/2
Ovary	–/1	1/1	–/1	–/1	1/1	1/1	
Astrocytoma	–/3	–/3	–/3	3/3	3/3	–/3	6/16
Melanoma	–/6	–/6	–/6	–/6	10/10	–/2	4/10
Lymphoblastoid cells	–/7	–/7	–/7	–/7	7/7		–/4
Normal human cells							
Kidney epithel.	25/25	25/25	25/25	25/25	10/10	15/15	–/4
Fetal kidney	2/3	3/3	–/3	3/3			
Adult fibrobl.	–/1	–/1	–/1	1/1	1/1		–
Fetal fibrobl.	–/1	–/1	–/1	1/1	1/1		–
Fetal brain	–	–	–	–	1/1		–
Erythrocytes	–	–	–	–	–		–

lysis appears in table II. Detailed specificity may be obtained from the original references as noted in table I.

Representative immunohistologic photomicrographs of normal adult kidney are shown in figure 2. A summary of the tissue section specificity for these mAb within the kidney is shown in figure 3; mAb S 22 (gp115) could not be identified in any normal kidney cells or any normal cells outside the kidney.

The tissue distribution of these antigens has also been studied on nonrenal tissue. The data is beyond the scope of this discussion and

Fig. 2. Immunofluorescence *(A–F)* and immunoperoxidase *(G, H)* photomicrographs of normal kidney sections: *(A)* mAb A99 localizes to glomerulus (×200); *(B)* mAb S4 localizes to glomerulus and proximal tubules. Distal tubules are not stained (×200). mAb AJ8 provides a similar staining pattern. *(C)* mAb S27 localizes to the proximal tubule as it exits Bowman's capsule (×400). *(D)* mAb S27 on the proximal portion of the loop of Henle (×200). mAb S23 demonstrates a similar pattern. *(E)* mAb C26 localizes to distal tubules (×200). *(F)* mAb C26 on collecting tubules (×200, 400). *(G)* mAb S8 localizes to endothelial cells (×400). Proximal and distal tubular epithelium are negative. *(H)* mAb S8 localizes to the collecting tubule especially at the luminal surface (×400).

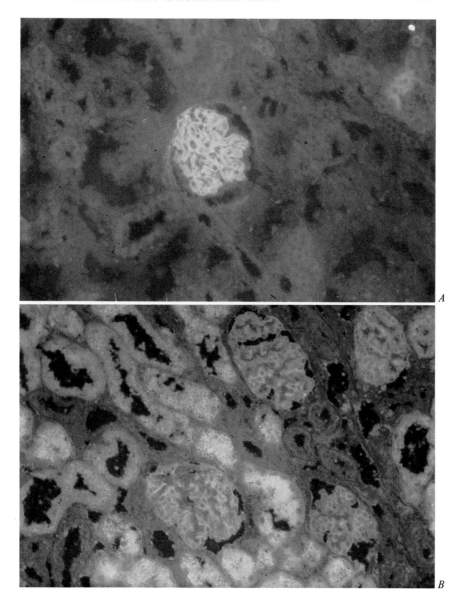

will be published elsewhere [11]. It is of particular interest and importance, however, that the immunohistologic distribution of all 9 of these defined antigens very closely resembles the specificity pattern obtained in vitro and, therefore, lends support to the validity and util-

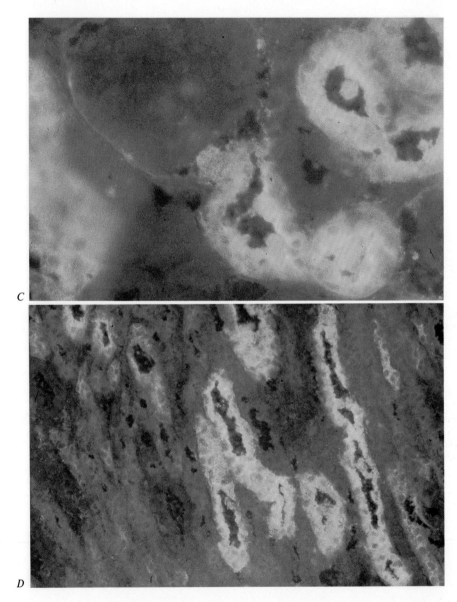

ity of the study of tissue culture lines. This is not to imply that we feel in vitro lines are identical to their in vivo cellular counterparts but rather that they provide reasonable approximations.

The data in figure 3 provide a point of reference. That is, they

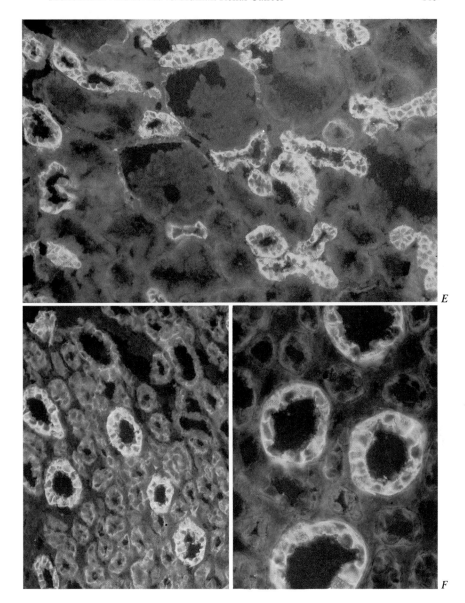

provide the antigenic phenotype of the various cells comprising the nephron. For example, proximal tubular cells are A99$^-$/AJ8$^+$/gp160$^+$/gp120r$^+$/gp120nr$^+$/ABH$^-$/gp40$^-$, distal tubular cells are A99$^-$/AJ8$^-$/gp160$^-$/gp120r$^-$/gp120nr$^-$/ABH$^-$/gp40$^+$ and

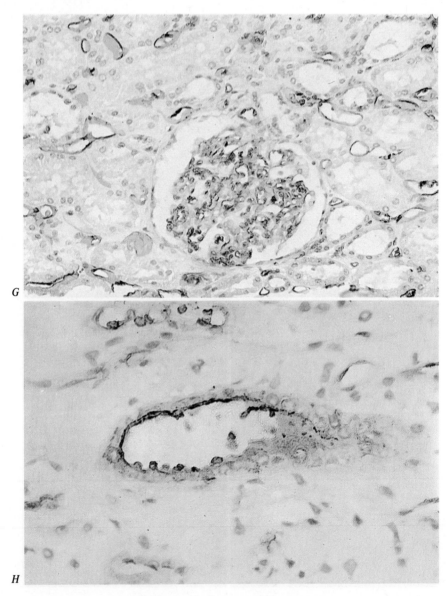

collecting tubular cells are $A99^-/AJ8^-/gp160^-/gp120r^-/gp120nr^-/ABH^+/gp40^+$.

When normal kidney cells in tissue culture are studied with these mAb, it becomes apparent that the morphologically heterogeneous

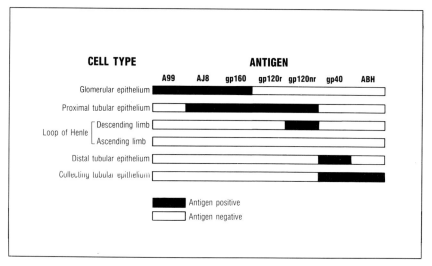

Fig. 3. Antigenic expression of normal human kidney cells defined by immunofluorescence and immunoperoxidase localization of respective mAb on frozen and/or paraffin tissue sections.

cells are derived from different sites in the nephron. Approximately 70% of the cells express a phenotype consistent with proximal tubular derivation. The remaining cells are mixtures of distal and collecting tubular and glomerular cells.

When frozen sections of 32 renal cancers are immunopathologically studied with these mAb, 31 (97%) typed $A99^-/gp120nr^+/ABH^-/gp40^-$. This phenotype is consistent with that of proximal tubular cells. Indeed, earlier immunologic studies with heteroantisera [12] and ultrastructural studies [13, 14] had already implicated proximal tubular cells as the source of renal carcinoma.

These 31 renal cancers were also studied for expression of AJ8, gp160, gp120r and gp115. The antigen defined by mAb AJ8, although immunohistologically present on all normal proximal tubular cells, could not be demonstrated on any renal carcinomas. Conversely, gp115, which could not be immunohistologically demonstrated on any normal cells, can be detected on 12 of 31 renal cancer specimens. The significance of these two interesting observations requires further investigation. Perhaps these represent examples of repression and depression, respectively, of gene product expression concomitant with neoplastic transformation.

Expression of gp160 and gp120r, like gp115, was variable. That is, a given tumour might express some, none or all of these three antigens. This was in fact, just what was observed in vitro [7]. When the 3 Ag (gp 160, gp120r, gp115) phenotype of these tumours was determined, it was anticipated that results might be analogous to those previously reported in lymphomas/leukemias and melanomas [15]; that is, coordinate expression of certain antigens thereby deriving patterns corresponding with different stages of differentiation. This was not the case.

Of 8 possible phenotypic patterns, all 8 were seen (fig. 4). Furthermore, analysis of the distribution into each subset approached that expected if the 3 Ags were expressed independently, or incoordinately, of each other.

At least two explanations for this observation are possible. One may be seeing loss of previously expressed differentiation Ags in the process of neoplastic transformation or, perhaps, development of carcinoma from cells which never normally differentiated. The independent, incoordinate nature of expression of these Ags makes it less likely that these neoplasms arise at points along the differentiation

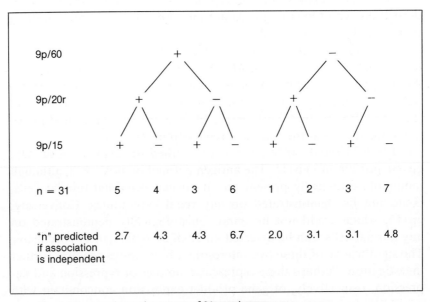

Fig. 4. Three antigen phenotype of 31 renal cancers.

pathway of proximal tubular cells. Of the 3 tumour types in which cell surface Ags have been phenotypically characterized, that is leukemia/lymphoma, melanoma and renal cancer, the latter differs from the former two which appear to derive at identifiable points along their respective differentiation pathways. It appears, therefore, that renal cancer may offer a model of neoplasia different from lymphomas or melanomas.

Comparison of antigenic phenotype to clinical parameters reveals a relationship between antigen expression and the pathological stage and grade of the tumour. Those tumours which most closely express a full complement of differentiation antigens tend to be low grade/low stage. At the other end of the spectrum, tumours expressing none of these three antigens were high grade/high stage lesions. Partial expression of these antigens correlated with moderate differentiation and intermediate stage. This observation is well exemplified in the case of gp120r expression (fig. 5). Six of 7 (86%) stage I tumours were gp120r$^+$ while 9 of 10 (90%) stage IV tumours were gp120r$^-$. Of 14 stage II and III patients, 5 were gp120r$^+$ and 9 were gp120r$^-$. A prospective study is currently underway to look at whether antigenic phenotype correlates with prognosis. It will be of interest, for in-

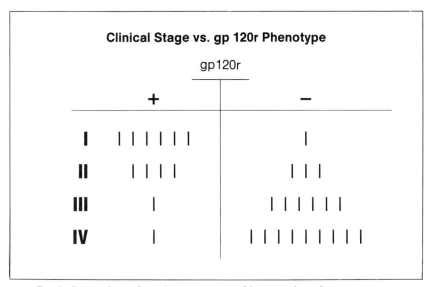

Fig. 5. Comparison of renal tumour stage with expression of gp120r.

stance, to see if gp120r typing will segregate patients into low or high risk subsets and whether it will prove to be a more accurate predictor of outcome than the currently available criteria of stage and grade.

Conclusions

Understanding human cancer cannot rely solely on the study of cancer in laboratory animals. Yet logistic and ethical considerations impose severe limits on human experimentation. Further progress will require the study of human cancers themselves. This may now be approached efficiently and effectively with the use of human cancer cell lines.

Renal cancer offers a highly useful tumour model in vitro. These tumours can be immortalized with a relatively high success rate while the autologous normal kidney epithelium can grow in short term cultures.

mAb analysis of human kidney antigens has begun to define the antigenic phenotype of the different cells which comprise the nephron. Comparison of the cell surface antigenic phenotype of in vitro lines and their in vivo counterparts reveal a marked, though not absolute, correspondence which supports the validity of studying in vitro models. Additionally, this data confirms earlier work establishing the origin of renal cancer from proximal tubular cells. Proximal tubular cells can be shown to make up about 70% of the cells in normal kidney cultures. One, therefore, has available a unique system with viable human cancer cells and their autologous normal cellular counterparts.

Futhermore, mAb probes demonstrate that human renal cancers do not express a full complement of normal proximal tubular differentiation antigens. The pattern of antigenic expression of renal cancers is unlike the selective patterns seen in lymphomas or melanomas indicating the point along the differentiation pathway where these latter cancers arise. Rather, in renal cancer, differentiation antigens are expressed or repressed independently of one another. This probably represents independent loss of antigens in the transformation process or development of cancer in a cell which had never normally differentiated. Regardless of which explanation proves correct, renal cancer appears to offer a model system different from others currently under investigation.

In addition, preliminary study indicates that antigenic phenotype may correlate with the malignant potential of the individual tumour.

The availability of the renal cancer system promises further insight into human cancer.

Acknowledgments: The author wishes to thank Drs. *J. Mattes, J. Sakomoto* and *G. Cairncross* for the gifts of mAb A99, C26 and AJ8, respectively. In addition, *Rosemarie Ramsawak, Connie Finstad* and Drs. *C. Cordon-Cardo, W. F. Whitmore, Jr., H. F. Oettgen,* and *L. J. Old* provided invaluable help.

This work was supported by an NIH Immunobiology Training Grant, Ferdinand C. Valentine Fellowship, and the Memorial Hospital Genitourinary Fund.

References

1 Gross, L.: Intradermal immunization of C3H mice against a sarcoma that originated in an animal of the same line. Cancer Res. *3:* 326–333 (1943).
2 Foley, E. J.: Antigenic properties of methylcholanthrene induced tumors in mice of the strain of origin. Cancer Res. *13:* 835–837 (1953).
3 Prehn, R. T.; Main, J. M.: Immunity to methylcholanthrene induced sarcomas. J. natn. Cancer Inst. *18:* 769–778 (1957).
4 Old, L. J.: Cancer immunology: the search for specificity-G.H.A. Clowes Memorial Lecture. Cancer Res. *41:* 361–375 (1981).
5 Bander, N. H.: Comparison of antigenic expression of human renal cancers in vivo and in vitro. (Cancer, in press).
6 Ueda, R., Shiku, H.; Pfreudshuh, M.; Takahashi, T.; Li, L.; Whitmore, W. F.; Oettgen, H.; Old, L. T.: Cell surface antigens of human renal cancer defined by autologous typing. J. exp. Med. *150:* 564–579 (1979).
7 Ueda, R.; Ogata, S.-I.; Morrissey, D. M.; Finstad, C. L.; Szkudlarek, J.; Whitmore, W. F.; Oettgen, H. F.; Lloyd, K. O.; Old, L. J.: Cell surface antigens of human renal cancer defined by mouse monoclonal antibodies: Identification of tissue-specific kidney glycoproteins. Proc. natn. Acad. Sci. USA *78:* 5122–5126 (1981).
8 Mattes, M. J.; Cairncross, J. G.; Old, L. J.; Lloyd, K. O.: Monoclonal antibodies to three widely distributed human cell surface antigens. Hybridoma (in press).
9 Sakomoto, J.; Cordon-Cardo, C.; Friedman, E.; Finstad, C.; Enker, W.; Melamed, M.; Lloyd, K.; Oettgen, H. F.; Old, L. J.: Antigens of normal and neoplastic human colonic mucosa cells defined by monoclonal antibodies (Abstract No. 889). Am. Ass. Cancer Res. (1983).
10 Cairncross, J. G.; Mattes, M. J.; Beresford, H. R.; Albino, A. P.; Houghton, A. N.; Lloyd, K. O.; Old, L. J.: Cell surface antigens of human astrocytoma defined by mouse monoclonal antibodies: identification of astrocytoma subsets Proc. natn. Acad. Sci. USA *79:* 5641–5645 (1982).
11 Finstad, C. L.; Bander, N. H.; Cordon-Cardo, C. C.; Whitmore, W. F.; Oettgen,

H. F.; Old, L. J.: In vivo specificity of mouse monoclonal antibodies to human renal cancer antigens. (Manuscript in preparation).
12 Wallace, A. C.; Nairn, R. C.: Renal tubular antigens in kidney tumors. Cancer *29:* 977 (1972).
13 Seljelid, R.; Ericsson, L. E.: Electron microscopic observations of the cell surface in renal clear cell carcinoma. Lab. Invest. *14:* 435 (1965).
14 Tannenbaum, M.: Ultrastructural pathology of human renal cell tumors. Pathol. A. *6:* 249 (1971).
15 Houghton, A. N.; Eisinger, M.; Albino, A. P.; Cairncross, J. G.; Old, L. J.: Surface antigens of melanocytes and melanomas: markers of melanocyte differentiation and melanoma subsets. J. exp. Med. *156:* 755–766, (1982).

N. H. Bander, M. D., Dept. of Surgery – Urology, Laboratory of Human Cancer Serology, Memorial Sloan-Kettering Cancer Center, 1275 York Avenue, New York, N. Y. 10021 (USA)

Three Human Melanoma-Associated Antigens and Their Possible Clinical Application

K. E. Hellström[1, 2], I. Hellström[1, 3], J. P. Brown[1, 2], S. M. Larson[4, 5], G. T. Nepom[1, 2], J. A. Carrasquillo[5]

[1] Program of Tumor Immunology, Fred Hutchinson Cancer Research Center, Seattle, Wa., USA
[2] Departments of Pathology, [3] Microbiology/Immunology, [4] Laboratory Medicine, Nuclear Medicine, and Radiology, University of Washington, Seattle, Wa., USA
[5] Seattle Veterans Administration Medical Center, Seattle, Wa., USA

Introduction

Melanoma is one of the human neoplasms that have been intensely studied by immunological techniques [1, 2]. Autologous serum antibodies have been detected to antigens that are unique to a given patient's tumour, as well as to antigens shared by most melanomas [3], and cell-mediated immune responses have been demonstrated to melanoma-associated antigens [4–6].

Over the past 4–5 years, many monoclonal mouse antibodies have been raised against cell surface antigens that are expressed in larger amounts by melanomas than by normal tissues [7–9]. Some of these melanoma-associated antigens have sufficient specificity to be of potential clinical interest for diagnosis and therapy. We shall summarize work performed in our laboratory on 3 melanoma-associated antigens which fall into this category: a glycoprotein (p97), a proteoglycan, and a sialoganglioside.

Nature of the Three Antigens

p97: p97 is a sialoglycoprotein first described by our group [10]. Its molecular weight (MW) on sodium dodecyl sulfate polyacrylamide gel electrophoresis *(SDS-PAGE)* is slightly smaller than that of rabbit phosphorylase b (97,400). An antigen, gp95, described by

Dippold et al. [11], has been proven to be the same as p97 [12], as has an antigen [13] described by Dent's group (*Brown,* unpublished observations). Monoclonal antibodies have been made to 5 different epitopes of p97, three of which are present on a 40,000 MW fragment obtained by proteolytic digestion [12].

According to *N*-terminal amino acid sequencing, p97 is related to transferrin [14]. Like transferrin, p97 binds iron [14]. The antigen is coded for by a gene that has been assigned to chromosome 3 [15], the same chromosome that carries the gene for the transferrin receptor [16], and probably also the gene for transferrin [17]. Recently, *Brown et al.* have cloned cDNA sequences coding for p97 [18].

A quantitative study of p97 expression in various tissues has been performed with binding assays [19]. According to this study, p97 is most strongly expressed in melanomas derived either from culture or from biopsy material, so that approximately 50% of melanomas express 50,000–400,000 molecules of p97 per cell. Occasional tumours other than melanoma have relatively high levels of p97 (up to 50,000 molecules per cell), as do nevi and foetal intestine. Normal adult tissues also contain p97 but at much lower levels (less than 8,000 molecules per cell), the highest amounts being detected in smooth muscle. Bone marrow cells and leukocytes express very low levels of p97 (less than 1,000 molecules per cell).

The results of the antibody-binding studies were confirmed and extended by an immunohistological investigation, which demonstrated p97 in biopsies of most primary and metastatic melanomas [20] and in nevi. A weak expression of p97 was detected in myoepithelial cells [20] and in cells from liver parenchyma (*I. Hellström,* unpublished observations). We shall comment further on the immunohistological data below.

Nepom et al. have recently immunized rabbits with a monoclonal anti-p97 antibody, 8.2, specific for p97^c, one of the 5 identified epitopes of p97, and obtained an antiserum, which after extensive absorption has the characteristics of an idiotype-specific antibody [21]. This antibody binds several different mouse antibodies to p97^c, but not to antibodies directed against other epitopes of p97 or to control antigens. The binding of the absorbed rabbit antiserum to antibody 8.2 can be competitively inhibited by p97. Mice injected with the rabbit antiserum display delayed-type hypersensitivity *(DTH)* to a human tumour line expressing 400,000 molecules of p97 per cell but not

Table I. Characteristics of anti-idiotypic antibodies (anti-Id) made to a p97-specific monoclonal antibody, 8.2 (Id)[a]

I. Specificity

    A. Anti-Id bind monoclonal antibody 8.2, an anti-p97^c IgG1 murine immunoglobulin.

    B. Binding of anti-Id to 8.2 is partially inhibited by
        1. soluble p97 antigen;
        2. 3 other IgG1 anti-p97 monoclonal antibodies, but not 7 other anti-p97 antibodies with other isotypes or epitope specificity.

II. In vivo activity

    A. Mice immunized with anti-Id demonstrate DTH reactivity to p97-containing human melanomas cells.

    B. Mice immunized with anti-Id make immunoglobulin
        1. which bear the 8.2 idiotype;
        2. some of which bind soluble p97 antigen.

[a] From *Nepom et al.*, 1983 [21]

to a tumour line expressing only 2,000 molecules of p97 per cell, and some of the immunized mice have circulating antibodies which bind to the rabbit antiserum and which can immunoprecipitate p97. We conclude that the rabbit antiserum contains anti-idiotypic antibody capable of inducing an immune response to p97. The data are summarized in table I.

GD3 Ganglioside: Dippold et al. [11] have described an antibody, R_{24}, to a glycolipid antigen that is strongly expressed on most melanomas and only weakly, or not at all, by other tumours and normal adult tissues. This antigen was shown to be GD3 sialoganglioside [22]. Our group obtained a monoclonal IgM antibody, 4.2, to a similar antigen [23], also found to be GD3 sialoganglioside, which, however, differ from GD3 of normal brain in having longer fatty acids [24].

We have recently obtained several IgG1 antibodies to the GD3 antigen. One of these antibodies, 2B2, gives high antibody-dependent cellular cytotoxicity (*ADCC; I. Hellström*, unpublished data), even though it is IgG1, and antibody which is normally less effective than IgG2 in giving ADCC [25]. Antibody 2B2 has a very high specificity for melanoma when tested with immunohistological techniques; while it stained 96% of melanomas, we have only detected staining at a very low level of some normal adult tissues, including brain, and only very few tumours of other origin (table II and I. *Hellström*, unpublished data).

Proteoglycan Antigen: Bumol and Reisfeld [26] and *Imai et al.* [27] were the first to describe monoclonal antibodies to antigen(s) detected, on SDS-PAGE, as molecules of approximately 250,000 MW. The antigen detected by *Bumol and Reisfeld* was found to be a proteoglycan [26]; whether or not the antigen described by *Imai et al.* is identical is not yet clear.

We have obtained a monoclonal antibody, 48.7 [28], to a similar antigen expressed by most melanomas and, in smaller amounts, by some carcinomas. Normal tissues do not show appreciable amounts of this antigen, according to immunohistological studies, except for some staining of occasional endothelial (?) cells of the blood vessel wall [28]. Recently, we have obtained several additional antibodies to the antigen defined by antibody 48.7, which are specific for epitope(s) different from that defined by antibody 48.7.

Diagnostic Use of Antibodies to the Three Melanoma Antigens

In Vitro

Immunohistological studies are summarized in tables II–IV. They showed that antibodies to p97, the GD3, and the proteoglycan antigens stained melanoma cells from tumour biopsies while they stained cells from other tumours much less frequently. Approximately 55% of melanomas, as compared to 0% of other tumours (mostly

Table II. Summary of immunohistology data: PAP staining of frozen sections by each antibody

Designation	Antibody Specific for	≥2 + Staining[a]	
		Melanoma	Other tumours
96.5	p97	44/57 (77%)	10/35 (29%)
48.7	Proteoglycan antigen	50/57 (88%)	6/35 (17%)
2B2	GD3 Melanoma antigen	26/27 (96%)	1/15 (7%)

[a] Degree of reactivity graded from + (weak staining of most melanoma cells) to 4+ (very strong staining of most melanoma cells). Non-tumour cells in sections were not stained

Table III. Summary of immunohistological data: Staining for 2 of the 3 melanoma antigens[a]

Melanomas	26/27 (96%)
Other tumours	3/15 (20%)

[a] See footnote to table II

carcinomas) were stained by all three antibodies (table IV), while 96% of melanomas and 20% of other tumours stained by at least two of the antibodies (table III) of which 2B2 was the most melanoma specific (table II).

The degree of discrimination between melanoma and other tumours is not sufficient for us to recommend the use of antibodies to the three antigens for the routine diagnosis of melanoma. Clinically important information is, nevertheless, obtained by using the antibodies. First, the immunohistological tests suggest which antibodies to select for in vivo tumour diagnosis by radioimaging and therapy (see further below). Second, they may, when combined with regular histological studies, facilitate the diagnosis of metastatic tumours of unknown origin, particularly since lymphomas and sarcomas have not stained for any of the three melanoma antigens. Since nevi were stained by all three antibodies, differential diagnosis between malignant melanoma and nevi can obviously not be made by using these antibodies.

In Vivo

Imaging of tumours by using radiolabelled antibodies has attracted much attention [29, 30], since it may provide prognostic information by indicating the amount of tumour spread and also since it can give evidence for antibody uptake by tumour as a basis for therapy. Monoclonal antibodies are preferable to conventional antisera for this purpose.

We have used both whole monoclonal antibodies to p97 [31], and Fab fragments [32, 33] prepared from antibodies specific for either p97 or the proteoglycan antigen, to image metastatic melanoma in vivo. Antibodies were labelled with ^{131}I, except in one case when we

Table IV. Summary of histological data: ≥ Staining for 3 of the 3 melanoma antigens[a]

Melanomas	15/27 (55%)
Other tumours	0/15 (0%)

[a] See footnote to table II

used ^{123}I to perform single-photon tomography, which gave better resolution [32]. By using whole anti-p97 antibody to image patients with melanomas expressing moderate-to-high levels of p97, 88% of clinically detectable lesions were observed; the lesions not observed were primarily those that were less than 1.5 cm in diameter. In several cases, previously undetected tumour metastases were seen by imaging.

There are three kinds of data indicating that the imaging was tumour-antigen specific. First, experiments using differently labelled (^{131}I and ^{125}I) specific and control immunoglobulin preparations showed preferential uptake of the specific preparations in tumour and about equal uptake in normal tissues [31]. Second, when the same patients were imaged repeatedly, using both specific and control antibody preparations, only the specific ones showed uptake in tumour [32]. Third, a correlation was found between antibody uptake in tumour and the degree of antigen expression there [33].

Radiolabelled Fab fragments specific for the proteoglycan antigen have given at least as good localization in tumour as Fab specific for p97 (*Larson et al.,* unpublished findings). It is noteworthy that the Fab specific for the proteoglycan antigen do not localize in normal liver, while Fab (and whole antibody) specific for p97 do; the latter is probably due to the greater expression of p97 in liver (*I. Hellström,* unpublished findings).

Our data suggests that the imaging technique already gives clinically useful information, and further improvements of this technique may provide a diagnostically very powerful tool. The use of isotopes that can provide better resolution than ^{131}I for labelling may help toward this, for example, ^{96m}Tc or ^{111}In.

One complication to the imaging work is that patients produce antibodies to mouse immunoglobulins. These antibodies occur almost always in patients given whole immunoglobulin, and more sporadically and after more injections in patients given Fab fragments.

Therapy

The imaging experiments summarized in the previous section showed that radiolabelled monoclonal antibodies (or Fab) specific for p97 or the proteoglycan antigen localize in metastatic melanoma when injected intravenously into patients. Further evidence for such localization comes from studies in which patients were injected intravenously with a mixture of anti-p97 and anti-proteoglycan antibodies, 212 mg of each, over a 10-day period. Melanoma metastases were removed the day of the last injection and were assessed by immunohistology for the presence of antibody bound to the tumour tissue. Antibody was observed on the melanoma cells in the sections, while the surrounding stroma was negative.

In view of the evidence that anti-melanoma antibodies localize into melanoma tissue in vivo, such antibodies should be useful for carrying tumour-destructive agents to the site of a growing neoplasm. As a first step towards this objective, we have used Fab fragments specific for either p97 or the proteoglycan antigen, after labelling with ^{131}I, to carry a potentially therapeutic dose of radiation into the sites of metastic melanoma [32, 33]. Dosimetric studies indicated that it would be possible, using such an approach, to bring approximately 1000 rads to the tumour site for each 35 rads being delivered to the bone marrow, the most radiosensitive of the normal sites [33].

Based on the dosimetric studies, a Phase I therapeutic trial was initiated in 9 patients with widespread metastatic melanoma [33], giving ^{131}I-labelled Fab fragments in doses up to 100 mCi (7 patients), 400 mCi (1 patient), and 850 mCi (1 patient). The two patients receiving the highest doses both responded to therapy; in one patient there was an 8-month arrest to tumour growth and in the other patient there was a complete remission of some metastatic nodules and partial remission of others. Notably, none of the patients on the Phase I study had any severe reactions. For example, the two patients receiving the highest doses were able to continue working except for the short period that they stayed in the hospital for injection of the radiolabelled antibody fragments.

In addition to using antibodies to deliver a therapeutic dose of radiation to tumours, effort should go into selecting antibodies which can mediate strong ADCC [25] and, therefore, be therapeutically useful by themselves. Conjugates between antibodies and toxins [34] or chemotherapeutic drugs [35, 36] are also likely to be useful for providing specific therapy with selective toxicity to tumour cells expressing moderate to high levels of the target antigens. A currently pressing problem to solve for future therapeutic as well as diagnostic work with antibodies (and Fab fragments), is to develop ways for decreasing the formation of antibodies, in patients, to the injected mouse immunoglobulin preparations. The appearance of such antibodies in patients leads to immediate inactivation of the injected immunoglobulins and could conceivably give other complications as well.

As a different approach to therapy, one should consider the use of specific immunogens in the form of purifed tumour antigens or synthetic peptides [37], using, as the therapeutic targets, tumour markers that have been first identified by monoclonal antibodies. Anti-idiotypic antibodies [21, 38] may be similarly useful as immunogens, as illustrated by the work summarized above.

Summary

Monoclonal antibodies have been made to three different cell surface antigens (p97, a proteoglycan, a GD3 sialoganglioside) expressed most strongly by human melanoma, and used to characterize the antigens with respect to molecular nature and degree of expression by various cells in vivo. Radiolabelled antibody fragments specific for two of the antigens (p97 and the proteoglycan) have proven diagnostically useful for tumour imaging in patients with metastatic melanoma. A Phase I therapeutic trial with ^{131}I-labelled antibody fragments has been initiated with encouraging results.

Acknowledgments: The work of the authors was supported by Grants CA 19148, CA 19149, CA 29639, and CA 34777 from the National Institutes of Health; and Grant IM 241B and RD 154 from the American Cancer Society.

References

1 Hellström, K. E.; Hellström, I.: Immunity to neuroblastomas and melanomas. Annu. Rev. Med. 23: 19–38 (1972).
2 Reisfeld, R. A.; Ferrone, S.: Melanoma antigens and antibodies, pp. 1–433 (Plenum Press, New York, London 1982).

3 Shiku, H. J.; Takahashi, T.; Resnick, L. A.; Oettgen, H. F.; Old, L. J.: Cell surface antigens of human malignant melanoma III. Recognition of autoantibodies with unusual characteristics. J. exp. Med. 145: 784–789 (1977).
4 Hellström, I.; Hellström, K. E.; Sjögren, H. O.; Warner, G. A.: Demonstration of cell-mediated immunity to human neoplasms of various histological types. Int. J. Cancer 7: 1–16 (1971).
5 Herberman, R. B.: Cell-mediated immunity to tumour cells. Adv. Cancer Res. 19: 207–263 (1974).
6 Halliday, W. J.; Maluish, A. E.; Little, J. H.; Davis, N. C.: Leukocyte adherence inhibition and specific immunoreactivity in malignant melanoma. Int. J. Cancer 16: 645–658 (1975).
7 Koprowski, H.; Steplewski, Z.: Human solid tumour antigens defined by monoclonal antibodies; in Hämmerling, Hämmerling, Kearney, Monoclonal antibodies and T-cell hybridomas, perspectives and technical advances; Research monographs in immunology, No. 3, pp. 161–173 (Elsevier/North-Holland Biomedical Press, Amsterdam 1981).
8 Hellström, K. E.; Hellström, I.; Brown, J. P.: Monoclonal antibodies to melanoma-associated antigens; in Wright, Monoclonal antibodies and cancer (Marcel Dekker, New York, in press 1983).
9 Reisfeld, R. A.: Monoclonal antibodies to human malignant melanoma. Nature 298: 325–326 (1982).
10 Woodbury, R. G.; Brown, J. P.; Yeh, M-Y.; Hellström, I.; Hellström, K. E.: Identification of a cell surface protein, p97, in human melanomas and certain other neoplasms. Proc. natn. Acad. Sci. 77: 2183–2186 (1980).
11 Dippold, W. G.; Lloyd, K. O.; Li, L. T. C.; Ikeda, H.; Oettgen, H. F.; Old, L. J.: Cell surface antigens of human malignant melanoma: definition of six antigenic systems with mouse monoclonal antibodies. Proc. natn. Acad. Sci. 77: 6114–6118 (1980).
12 Brown, J. P.; Nishiyama, K.; Hellström, I.; Hellström, K. E.: Structural characterization of human melanoma-associated antigen p97 using monoclonal antibodies. J. Immunol. 127: 539–546 (1981).
13 Liao, S.-K.; Clarke, B. J.; Khosravi, M.; Kwong, P. C.; Brickenden, A.; Dent, P. B.: Human melanoma-specific oncofetal antigen defined by a mouse monoclonal antibody. Int. J. Cancer 30: 573–580 (1982).
14 Brown, J. P.; Hewick, R. M.; Hellström, I.; Hellström, K. E.; Doolittle, R. F.; Dreyer, W. J.: Human melanoma-associated antigen p97 is structurally and functionally related to transferrin. Nature 296: 171–173 (1982).
15 Plowman, G. D.; Brown, J. P.; Enns, C. A.; Schröder, J.; Nikinmaa, B.; Sussman, H. H.; Hellström, K. E.; Hellström, I.: Assignment of the gene for human melanoma-associated antigen p97 to chromosome 3. Nature 303: 70–72 (1983).
16 Enns, C. A.; Suomalainen, H. A.; Gebhardt, J. E.; Schröder, J.; Sussman, H. H.: Human transferrin receptor: Expression of the receptor is assigned to chromosome 3. Proc. natn. Acad. Sci. USA 79: 3241–3245 (1981).
17 McAlpine, P. J.; Bootsma, D.: Report of the committee on the genetic constitution of chromosomes 2, 3, 4, and 5. Cytogenet. Cell Genet. 32: 121–129 (1982).
18 Brown, J. P.; Rose, T. M.; Forstrom, J. W.; Hellström, I.; Hellström, K. E.: Isolation of a cDNA clone for human melanoma-associated antigen p97. (in preparation).
19 Brown, J. P.; Woodbury, R. G.; Hart, C. E.; Hellström, I.; Hellström, K. E.: Quantitative analysis of melanoma-associated antigen p97 in normal and neoplastic tissues. Proc. natn. Acad. Sci. USA 78: 539–543 (1981).
20 Garrigues, H. J.; Tilgen, W.; Hellström, I.; Franke, W.; Hellström, K. E.: Detec-

tion of a human melanoma-associated antigen, p97, in histological sections of primary human melanomas. Int. J. Cancer 29:511–515 (1982).
21 Nepom, G. T.; Nelson, K. A.; Holbeck, S. L.; Hellström, I.; Hellström, K. E.: Induction of immunity to a human tumour marker by in vivo administration of anti-idiotypic antibodies in mice. (in preparation).
22 Pukel, C. S.; Lloyd, K. O.; Trabassos, L. R.; Dippold, W. G.; Oettgen, H. F.; Old, L. J.: GD3, a prominient ganglioside of human melanoma. Detection and characterization by mouse monoclonal antibody. J. exp. Med. 155:1133–1147 (1982).
23 Yeh, M.-Y.; Hellström, I.; Abe, K.; Hakomori, S.; Hellström, K. E.: A cell surface antigen which is present in the ganglioside fraction and shared by human melanomas. Int. J. Cancer 29:269–275 (1982).
24 Nudelman, E.; Hakomori, S.; Kannagi, R.; Levery, S.; Yeh, M.-Y.; Hellström, K. E.; Hellström, I.: Characterization of a human melanoma-associated ganglioside antigen defined by a monoclonal antibody, 4.2. J. biol. Chem. 257: 12752–12756 (1982).
25 Hellström, I.; Hellström, K. E.; Yeh, M.-Y.: Lymphocyte-dependent antibodies to antigen 3.1, a cell surface antigen expressed by a subgroup of human melanomas. Int. J. Cancer. 27:281–285 (1981).
26 Bumol, T. F.; Reisfeld, R. A.: Unique glycoprotein-proteoglycan compex defined by monoclonal antibody on human melanoma cells. Proc. natn. Acad. Sci. USA 79:1245–1249 (1982).
27 Imai, K.; Wilson, B. S.; Kay, N. E.; Ferrone, S.: Monoclonal antibodies to human melanoma cells: comparisons of serological results of several laboratories and molecular profile of melanoma-associated antigens; in Hämmerling, Hämmerling, Kearney. Monoclonal antibodies and T-cell hybridomas, perspectives and technical advances; Research monographs in immunology, No. 3, pp. 183–190 (Else-vier/North-Holland Biomedical Press, Amsterdam, 1981).

Prof. Dr. K. E. Hellström, Oncogén, 3005 First Avenue, Seattle, WA 98121 (USA)

28 Hellström, I.; Garrigues, H. J.; Cabasco, L.; Mosely, G. H.; Brown, J. P.; Hellström, K. E.: Studies of a high molecular weight human melanoma-associated antigen. J. Immunol. 130:1467–1472 (1983).
29 Goldenberg, D. M.; DeLand, F.; Kim, E.; Bennett, S.; Primus, F. J.; Van Nagell, J. R.; Estes, N.; DeSimone, P.; Rayburn, P.: Use of radiolabelled antibodies to carcinoembryonic antigen for the detection and localization of diverse cancers by external photoscanning. New Engl. J. Med. 298:1384–1388 (1978).
30 Mach, J. P.; Buchegger, F.; Forni, M.; Ritschard, J.; Berche, C.; Lumbroso, J.-L.; Schreyer, M.; Girardet, C.; Accola, R. S.; Carrel, S.: Use of radiolabelled monoclonal anti-CEA antibodies for detection of human carcinomas by external photoscanning and tomoscintigraphy. Immunol. Today 2:239–249 (1981).
31 Larson, S. M.; Brown, J. P.; Wright, P. W.; Carrasquillo, J. A.; Hellström, I.; Hellström, K. E.: Imaging of melanoma with ^{131}I-labelled monoclonal antibodies. J. nucl. Med. 24:123–129 (1983).
32 Larson, S. M.; Carrasquillo, J. A.; Krohn, K. A.; McGuffin, R. W.; Hellström, I.; Hellström, K. E.; Lyster, D.: Diagnostic imaging of malignant melanoma with radiolabelled anti-tumour antibodies. J. Am. med. Ass. 249:811–812 (1983).
33 Larson, S. M.; Carrasquillo, J. A.; Krohn, K. A.; Brown, J. P.; McGuffin, R. W.; Ferens, J. M.; Graham, M. M.; Hill, L. D.; Beaumier, P. L.; Hellström, K. E.; Hellström, I.: Localization of p97 specific Fab fragments in human melanoma as a basis for radiotherapy. J. clin. Invest. (in press).
34 Casellas, P.; Brown, J. P.; Gros, O.; Gros, P.; Hellström, I.; Jansen, F. K.; Poncelet, P.; Vidal, H.; Hellström, K. E.: Human melanoma cells can be killed in vitro by an immunotoxin specific for melanoma-associated antigen p97. Int. J. Cancer 30:437–443 (1982).

35 Ghose, T.; Blair, A. H.: Antibody-linked cytotoxic agents in the treatment of cancer: current status and future prospects. J. natn. Cancer Inst. 61: 657–676 (1978).
36 Rowland, G. F.; Simmonds, R. G.; Corvalan, J. R. F.; Baldwin, R. W.; Brown, J. P.; Embleton, M. J.; Ford, C. H. J.; Hellström, K. E.; Hellström, I.; Kemshead, J. T.; Newman, C. E.; Woodhouse, C. S.: Monoclonal antibodies for targeted therapy with vindesine; in Hämmerling, Hämmerling, Kearney, Protides of the biological fluids. Proceedings of colloquium 30, pp. 375–379 (Pergamon Press, Oxford 1983).
37 Atassi, M. Z.: Precise determination of the entire antigenic structure of lysozyme: molecular features of protein antigenic structures and potential of surface stimulation synthesis – a powerful new concept for protein binding sites. Immunochemistry 15: 909–936 (1978).
38 Bluestone, J. A.; Sharrow, S. O.; Epstein, S. L.; Ozato, K.; Sachs, D. H.: Induction of anti-H-2 antibodies without alloantigen exposure by in vivo administration of anti-idiotype. Nature 291: 233–235 (1981).

Prof. Dr. K. E. Hellström, Oncogén, 3005 First Avenue, Seattle, WA 98121 (USA)

Monoclonal Antibodies Directed against Transformation-Related Antigens in Melanoma

J. P. Johnson[1], B. Holzmann[1], P. Kaudewitz[2], G. Riethmüller[1]

[1] Institute for Immunology, University of Munich, Munich, FRG
[2] Department of Dermatology, University of Munich, Munich, FRG

Introduction

The concept of tumour restricted antigens, i. e., structures which would qualitatively distinguish a malignant cell from its normal counterpart, is the foundation of tumour immunology. The evidence for the existence of such structures comes primarily from the demonstration of immune recognition and tumour rejection in UV and chemically induced murine tumours [1, 2]. Like minor histocompatibility antigens, these "tumour specific transplantation antigens" can only be detected by cellular immune responses. Recent analyses using cloned cytolytic T cells indicate an intriguing antigenic complexity in this system [3]. However, an understanding of the nature of tumour antigens requires that they can be isolated and biochemically defined. Attempts to detect such antigens with antibodies present in syngeneic or xenogeneic antisera have been generally unsuccessful due to the complexity and weak reactivities of the antibodies. Recent technological developments in two areas, recombinant DNA and monoclonal antibodies, open the possibility to search directly for the existence of genes and gene products which are present in transformed cells and not detectable in their normal counterparts.

Malignant melanoma was one of the first human tumours in which monoclonal antibodies were used in an attempt to define tumour restricted surface antigens. The first antibodies obtained were reported to detect melanoma restricted surface antigens when analysed on cell lines [4, 5] and tumour biopsies [6]. This conclusion was soon found to be overly optimistic, as a broadening of the test cell panel and the investigation of tissue sections revealed much wider reactivity. Extensive analyses of the antimelanoma monoclonal antibodies by ten laboratories participating in the NIH Melanoma Exchange Programme showed that most antibodies reacted with a variety of normal cells and structures [7]. However, several surface antigens strongly associated with melanoma cells have been defined. These include a proteoglycan [8], the GD3 ganglioside [9], and the transferrin related glycoprotein p97 [10]. In this paper, antimelanoma monoclonal antibodies produced and selected using different protocols are characterized for their reactivity with frozen tissue sections containing malignant or benign pigment cell lesions.

Materials and Methods

BALB/c or (BALB/c × C57Bl/6)F1 mice were immunized by a single i. p. injection of $5-7 \times 10^6$ cultured melanoma cells or fresh tumour cells. In fusion I, cell lines Mel Wei (2 injections) or Mel JuSo were used [11]. In fusions II and III lymph node metastases from 2 or 3 patients were pooled. *Bordetella pertussis* (fusions II, III) or directly coupled muramyl dipeptide (fusion III) were used as adjuvants. Spleens were removed three days after injection and fused with the myeloma P3 × 63 Ag 8.653 as described [11]. Growing hybrids were tested for antibody using a radiobinding assay (fusion I; [11]), a solid phase peroxidase linked ELISA (fusion II; [12]), or a single cell, beta-galactosidase linked ELISA (fusion III; [13]). Immunohistochemistry was performed on frozen sections using peroxidase coupled rabbit anti-mouse immunoglobulin and 3-amino-9-ethylcarbazol with H_2O_2 (*Holzmann et al.*, manuscript in preparation).

Results and Discussion

The power of the monoclonal antibody approach resides not only in the advantages of the reagents, but also in the potential to immortalize in vitro each antibody producing lymphocyte in the spleen. Thus given that the mouse is not tolerant, the specificities of the monoclonal antibodies eventually obtained after immunization with human tumour cells can be in large part a reflection of the immunization and screening procedures. The results from fusions using 3 different approaches to immunization and screening are summarized in table I.

Virtually all the reported monoclonal antibodies produced against melanoma have been obtained from mice immunized with

Table I. Approaches to the isolation of monoclonal antibodies against human melanoma

Fusion	I	II	III
Immunogen	Melanoma cell lines	Fresh tumor	Fresh tumour
Screening criteria	Immunizing cell line vs. Autologous PBMNC[a]	Pooled cell lines vs. Pooled PBMNC	Tumour cells vs. Pooled PBMNC
Selected	6/111[b]	32/371	27/257

[a] Peripheral blood mononuclear cells
[b] Number of hybrids selected/total number of hybrids

melanoma cell lines [14]. While cell lines are convenient sources of uncontaminated tumour cells, they may represent highly selected variants of the original tumour and express antigens more related to in vitro growth than to transformation. For these reasons, mice used in fusions II and III were immunized with melanoma cell from lymph node metastases. Antibodies in fusion I were selected on the basis of differential binding to melanoma cell lines and autologous peripheral blood mononuclear cells (PBMNC; [11]). In fusion II antibodies were selected on the basis of binding to a pool of melanoma cell lines but not to PBMNC pooled from 10 donors. Hybrids in fusion III were tested for binding to the immunizing tumour cells and to PBMNC using a single cell assay. This assay would enable antibodies directed to antigens present on a minor subpopulation of the tumour to be detected. As can be seen from table I, the fraction of monoclonal antibodies initially selected as defining melanoma related antigens ranged from 5.4% to 10.5% of all growing hybrids.

Melanoma is an ideal system to use in the search for transformation related markers. This tumour results from the malignant transformation of the pigment producing cells, the melanocytes. In the skin, melanocytes are found scattered throughout the epidermis along the dermal junction (figure 1). Proliferation of melanocytes gives rise to nevi which can be found in the epidermis (junctional nevus), the dermis (intradermal nevus), or in both (compound nevus). Although these lesions are benign, the cells often show characteristics of activated cells [14] and the majority of melanomas are reported to arise from melanocytes within nevi [15]. Thus, in this system it is possible to examine malignant melanocytes as well as benign cells from the same lineage simply by looking with immunohistochemical techniques at skin sections.

The monoclonal antibodies selected as shown in table I were examined on frozen sections of melanoma and nevi using an immunoperoxidase technique. Monoclonal antibodies directed against a monomorphic determinant on HLA-DR [16] and against p97 [10] were included in the screening.

Two antibodies were found which showed differential reactivity on melanoma and nevus (table II). Antibody 15.75 was obtained from a mouse immunized with a cell line and selected on the basis of differential binding to melanoma and autologous PBMNC (fusion I; [11]). Antibody P3.58 was obtained in fusion III in which fresh tumour cells were used both for immunization and screening. Table II presents the reactivity of these 2 antibodies together with anti p97 and anti HLA-DR on a panel of melanoma and nevi. As previously reported [17], p97 was expressed on cells in most of the nevi examined as well as in the melanomas. Although not demonstrable on in situ or

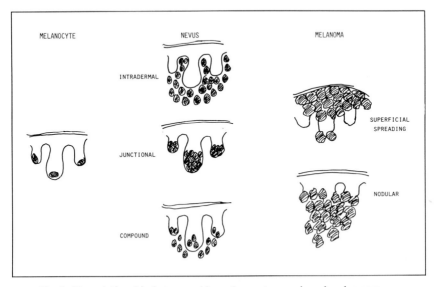

Fig. 1. The relationship between skin melanocytes, nevi, and melanoma.

Table II. Antigen expression on melanoma and nevi

	Antigen			
	15.75	P3.58	P97	HLA-DR
Melanoma	12/15[a]	13/15	9/10	15/15
Primary				
SSM[b]	5/8	6/8	6/7	8/8
NM[c]	4/4	4/4	1/1	4/4
Metastatic	3/3	3/3	2/2	3/3
Nevi	1/14	2/12	9/10	13/13
Intradermal	0/5	1/4	4/4	5/5
Compound	0/2	0/1	0/1	2/2
Junctional	1/4	1/4	3/3	3/3
Atypical nevus lentigo	0/1	0/1	NT[d]	1/1
Atypical melanocytic hyperplasia	0/1	0/1	1/1	1/1
Spitz nevus	0/1	0/1	1/1	1/1

[a] Number of positive samples/number tested.
[b] Superficial spreading melanoma.
[c] Nodular melanoma.
[d] Not tested.

in vitro proliferating melanocytes [18, 19], HLA-DR antigens were expressed on all of the melanomas and nevi. Expression of these antigens on both types of cells have been reported also by others [10, 21]. The majority of melanomas expressed the antigens defined by antibodies 15.75 and P3.58, and only 1 of the 15 examined was negative for both. In all cases the antibodies stained only a fraction of the tumour cells, and this remained constant in serial sections. In contrast, the majority of nevi were negative for both antigens. Of 12 nevi examined, only a single one (a junctional nevus) expressed both antigens.

The antigens defined by antibodies 15.75 and P3.58 are structurally distinct ([11]; *Holzmann, et al.,* manuscript in preparation) and distinct from the melanoma associated antigens thus far described [7]. Without exception, all melanoma associated antigens which have been examined have been shown to be expressed on melanocytes (in situ or cultured; [14, 19]) and/or on nevus cells [7, 14, 22]. The differences observed in melanoma and nevus expression of the 15.75 and P3.58 antigens may well be quantitative. Nevertheless, these antigens would appear to be the first described which demonstrate a transformation related change in the melanocyte cell lineage.

Acknowledgment: This work was supported in part by the Deutsche Forschungsgemeinschaft Bonn (SFB 37, Project B 9).

References

1 Prehn, R. T.; Main, J. M.: Immunity to methylcholanthrene-induced sarcomas. J. natn. Cancer Inst. *18:* 769 (1957).
2 Kripke, M. L.: Antigenicity of murine skin tumors induced by ultraviolet light. J. Natl. Cancer Inst. *53:* 1333 (1974).
3 Wortzel, R. D.; Phillips, C.; Schreiber, H.: Multiple tumor specific antigens expressed on a single tumor cell. Nature *304:* 165–167 (1983).
4 Koprowski, H.; Steplewski, Z.; Herlyn, D.; Herlyn, M.: Study of antibodies against human melanoma produced by somatic cell hybrids. Proc. natn. Acad. Sci. USA *75:* 3405–3409 (1978).
5 Yeh, M. Y.; Hellstrom, I.; Brown, J. P.; Warner, G. A.; Hansen, J. A.; Hellstrom, K. E.: Cell surface antigens of human melanoma identified by monoclonal antibodies. Proc. natn. Acad. Sci. USA *76:* 2927–2931 (1979).
6 Steplewski, Z.; Herlyn, M.; Herlyn, D.; Clark, W. H.; Koprowski, H.: Reactivity of monoclonal anti-melanoma antibodies with melanoma cells freshly isolated from primary and metastatic melanoma. Eur. J. Immunol. *9:* 94–96 (1979).
7 Steplewski, Z.: The second workshop on monoclonal antibodies to melanoma: Antigens of human melanoma as defined by monoclonal antibodies. Hybridoma *1:* 379–482 (1982).
8 Bumol, T. F.; Riesfeld, R. A.: Unique glycoprotein-proteoglycan complex defined by a monoclonal antibody on human melanoma cells. Proc. natn. Acad. Sci. USA *79:* 1245–1249 (1982).
9 Pukel, C. S.; Lloyd, K. O.; Travassos, L. R.; Dippold, W. G.; Oettgen, H. F.; Old, L. J.: GD3, a prominant ganglioside of human melanoma. Detection and characterization by a mouse monoclonal antibody. J. exp. Med. *155:* 1133–1147 (1982).
10 Brown, J. P.; Hewick, R. M.; Hellstrom, I.; Hellstrom, K. E.; Doolittle; R. F.; Dreyer, W. J.: Human melanoma-associated antigen p97 is structurally and functionally related to transferrin. Nature *296:* 171–173 (1982).
11 Johnson, J. P.; Demmer-Dieckmann, M.; Meo, T.; Hadam, M. R.; Riethmüller, G.: Surface antigens of human melanoma cells defined by monoclonal antibodies. I. Biochemical characterization of two antigens found on cell lines and fresh tumors of diverse tissue origin. Eur. J. Immunol. *11:* 825–831 (1981).
12 Johnson, J. P.; Riethmüller, G.: Tissue specificity of antimelanoma monoclonal antibodies analysed on cell lines. Hybridoma *1:* 381–386 (1982).
13 Holzmann, B.; Johnson, J. P.: A beta-galactosidase linked immunoassay for the analysis of antigens on individual cells. J. immunol. Methods *60:* 359–367 (1983).
14 Johnson, J. P.; Riethmüller, G.: The search for transformation related antigens in human tumors: the experience with monoclonal antibodies to melanoma; in Dierich, Ferrone, Handbook of monoclonal antibodies: application in biology and medicine. (Noyes Publications, Park Ridge, N. J., in press.)
15 McGovern, V. J.: Melanoma. Histological diagnosis and progress. (Raven Press, New York 1983.)
16 Herlyn, M.; Clark, W. H.; Mastrangelo, M. J.; Guerry, D.; Elder, D. E.; LaRossa, D.; Hamilton, R.; Bondi, E.; Tuthill, R.; Steplewski, Z.; Koprowski, H.: Specific immunoreactivity of hybridomasecreted monoclonal anti-melanoma antibodies to cultured cells and freshly derived human cells. Cancer Res. *40:* 3602–3608 (1980).

17 Garrigues, H. J.; Tilgen, W.; Hellstrom, I.; Franke, W.; Hellström, K. E.: Detection of a human melanoma associated antigen p97 in histological sections of primary human melanoma. Int. J. Cancer 29: 511–515 (1982).
18 Klareskog, L.; Malmnäs-Tjernlund, U.; Forsum, U.; Peterson, P. A.: Epidermal Langerhans cells express Ia antigens. Nature 268: 248 (1977).
19 Houghton, A. N.; Eisinger, M.; Albino, A. P.; Cairncross, J. G.; Old, L. J.: Surface antigens of melanocytes and melanomas. Markers of melanocyte differentiation and melanoma subsets. J. exp. Med. 156: 1755–1766 (1982).
20 Winchester, R. J.; Wang, C. Y.; Gibofsky, A., Kunkel, H. G.; Lloyd, K. O.; Old, L. J.: Expression of Ia like antigens on cultured human malignant melanoma cell lines. Proc. natn. Acad. Sci. USA 75: 6235–6239 (1978).
21 Thompson, J. J.; Herlyn, M. F.; Elder, D. E.; Clark, W. H.; Steplewski, Z.; Koprowski, H.: Expression of DR antigens in freshly frozen human tumors. Hybridoma 1: 161–168 (1982).
22 Herlyn, M.; Herlyn, D.; Elder, D. E.; Bondi, E.; LaRossa, D.; Hamilton, R.; Sears, H. F.; Balban, G.; Guerry, D.; Clark, W. H.; Koprowski, H.: Phenotypic characteristics of cells derived from precursors of human melanoma (manuscript submitted for publication).

Dr. Judith P. Johnson, Institut für Immunologie, Schillerstrasse 42,
D-8000 München 2 (FRG)

Comparative Analysis of Melanoma-Associated Antigens in Primary and Metastatic Tumour Tissue

J. Brüggen[1], E.-B. Bröcker[1], L. Suter[2], K. Redmann[1], C. Sorg[1]

[1] Department of Experimental Dermatology, University Clinic for Dermatology, Münster, FRG
[2] Fachklinik Hornheide, Münster-Handorf, FRG

Introduction

Human melanoma is a spontaneous tumour that progresses from an initial premalignant lesion to highly metastatic forms. Cells of primary and metastatic forms may differ in physiological and biochemical properties, which also includes the expression of surface constituents. We have already described our failure to detect melanomaspecific antigens; we found that human melanoma lines express a spectrum of melanoma-associated antigens which may also be expressed by normal cells and other tumours [1]. Moreover, we found that melanoma lines were heterogeneous, that is, they are composed of phenotypic variants, which could be isolated by cloning techniques [2]. The melanoma lines and their variant subclones have been characterized according to their biological, biochemical and structural properties [3–6]. This heterogeneity of melanoma cells in vitro reflects the in vivo situation, where melanomas reveal a heterogeneity not only with regard to growth characteristics but also to pigmentation and cell morphology. In a first attempt to type these differences serologically we used xenoantisera raised against melanoma lines in nonhuman primates; but due to limited specificity and titer of the sera the use in tumour tissue typing was restricted (*Terbrack et al.* unpublished observations).

Here we report the generation of monoclonal antibodies against human melanoma cell lines and melanoma biopsies, their specificity analysis on cell lines and sections of human melanoma of primary and metastatic origin.

Monoclonal Antibodies

Generation, Selection, and Reactivity with Cell Lines

For immunization of BALB/c mice we used melanoma lines, which differ with regard to biological and biochemical properties, such as plasminogen activator production, growth in the nude mouse, angiogenic activity and surface antigens [4]. The hybridoma supernatants were screened for antibodies using an enzyme linked immunosorbent assay (ELISA), which has been described before [7]. In the first stage supernatants were screened on pooled platelets and leukocytes; the negative ones were further tested in a second stage screening on various melanoma lines and, if consistently positive, tested in a third stage on purified pooled monocytes and B-lymphoid cell lines. Antibodies that were also negative at this third stage, were then analyzed for their fine specificity on various melanoma lines, their variant sublines and nonmelanoma tumour lines.

According to the reactivity on cell lines it was possible to define five groups of antigens, that were expressed only by melanoma lines, and to a minor degree, by embryonic fibroblasts (group I); by melanoma, neuroblastoma and teratoma lines (group II); by melanoma, neuroblastoma, teratoma, glioblastoma, carcinoma and embryonic fibroblasts (group III); by melanoma, teratocarcinoma and carcinoma (group IV); and in addition by B-lymphoblastoid cell lines (group V). The antibodies reacted qualitatively and quantitatively different with melanoma lines and their sublines [8].

Reactivity with Cryostat Sections of Tumour Biopsies

The antibodies characterized on various melanoma and nonmelanoma lines were then tested on cryostat sections of melanoma and nonmelanoma tumours or normal tissues. An indirect immunoperoxidase technique was applied, using 2-amino-ethylcarbazole (AEC) as chromogene and hemalaun and methylgreen as counterstain. Out of 34 monoclonal antibodies that bound to melanoma lines, only 19 antibodies reacted on melanoma tissues. Of the 19 antibodies 10 displayed broad cross-reactivity with other cells and structures, whereas nine had only restricted cross-reactivity. By their cross-reactivity with

Table I. Reaction patterns of monoclonal antibodies with normal structures in the skin

Type of antigen		Cross-reactivity
Nevocellular I	M-2-2-4	Single cells in sebaceous glands
Nevocellular II	M-2-7-6	Cutaneous nerves, weak
Neural	M-2-10-15	Cutaneous nerves
Endothelial	A-10-33	Capillaries
Basal cell	A-1-43	Basal cell layer of epidermis, sweat ducts, part of cutaneous nerves

normal structures and cells in vivo it was possible to define five categories of antibodies, that had no clear-cut relation to the reaction pattern of the antibody groups defined on cell lines (table I). Two antibodies were found which reacted with *nevocellular* antigens, but in different patterns, one with *neural,* one with *endothelium* and one with the *basal cell layer* of the epidermis. With other normal tissue and nonmelanoma tumours the antibodies show only restricted cross-reactivities [9].

Table II shows the frequency of reactivity with melanoma at different stages. Only the *nevocellular I* antigen is expressed in nearly 100% of the primary and metastatic melanomas and comes closest to the definition of a lineage-specific antigen. The *nevocellular II* and the *neutral* antigens occur to a different degree in various melanomas,

Table II. Reaction patterns of monoclonal antibodies with frozen sections of melanoma

Type of antigen	Melanomas primary	Metastases skin	Lymph nodes	Nevi
Nevocellular I M-2-2-4	92% 23/25	93% 42/45	88% 14/16	100% 18/18
Nevocellular II M-2-7-6	30% 10/33	20% 9/45	43% 6/14	40% 4/10
Neural M-2-10-15	26% 11/42	16% 7/45	50% 8/16	22% 4[a]/18
Endothelial A-10-33	22% 10/45	8% 3/40	50% 7/14	0/18
Basal cell A-1-43	35% 19/54	35% 17/49	69% 11/16	11% 2[a]/18
HLA-DR[b]	36% 29/80	31% 16/51	53% 8/15	0/9

[a] Weak reaction with less than 10% of cells; [b] MoAbs OKIa and 970D7

Table III. Biochemical characterization of melanoma associated antigens

Antigen		Molecular weight[a] u	Isotype
Nevocellular I	M-2-2-4	—[b]	IgG_1
Nevocellular II	M-2-7-6	–	IgG_1
Neural	M-2-10-15	200,000	IgG_1
Endothelial	A-10-33	–	IgG_1
Basal cell	A-1-43	130,000	IgG_1

[a] SDS-PAGE
[b] Not detectable by surface iodination

but no clear-cut association to a certain stage was obtained. The *endothelial* and the *basal cell* antigen are preferentially expressed in lymph node metastases. Both these antigens can be classified as "jumping antigens", which are expressed by other tissues but occur at certain stages during the tumour progression of malignant melanoma.

So far two antigens could be characterized biochemically: the *basal cell* antigen has a molecular weight of 130,000 u and the *neural* antigen of approximately 200,000 u. The other antigens are not detectable by surface iodination (table III).

In order to analyze the expression of melanoma-associated antigens in the course of tumour growth, we looked at the occurrence of these antigens in various types of melanoma. In addition to the melanoma-associated antigens described we included the HLA-DR antigen, which can be found on tumour cells within human malignant melanoma tissue [10].

Table IV. Expression of melanoma-associated antigens in different metastases

	Locoregional[a] n 32	Disseminated[b] n 30
M-2-2-4	93%	85%
M-2-7-6	49%	23%
M-2-10-15	39%	18%
A-10-33	20%	11%
A-1-43	67%	20%
HLA DR[c]	48%	30%

[a] Regionary lymph nodes and skin
[b] Skin and internal organs
[c] OKIa and 910D7

Table IV shows the distribution of these antigens in melanoma metastases. Locoregional metastases are compared to disseminated metastases. Apart from the M-2-2-4 antigen, which is found in approximately 90% of both types, the other antigens were found more frequently in locoregional than in distant metastases. This is particularly true for the A-1-43 antigen. Again the *nevocellular I* antigen is expressed up to 100%, whereas the other antigens show a more restricted distribution.

In table V the expression of the described antigens in primary tumours was investigated, with regard to their prognostic potential. The first three antigens occurred with the same frequency in primary tumours of patients, whether the patients developed metastases or not. The *endothelial, basal* cell and the *HLA-DR* antigens, however, had a significantly higher occurrence in primary tumours that had metastasized within the short observation period (mean 15.7 ± 7 months). A relation of the expression of these antigens to tumour thickness was also seen [11].

In another study concerning the changes of biochemical and antigenic properties of human melanoma lines, after treatment with biological response modifiers such as retinoic acid and Phorbolesters, it turned out that the *nevocellular I* antigen, which is expressed by nearly 100% of the cases tested, was not altered after treatment. It is interesting to note that the expression of more restricted antigens such as the *nevocellular II*, the *neural*, the *endothelial* and the *basal cell* antigens was significantly modulated (table VI), [12].

Table V. Antigen expression in primary cutaneous melanoma and occurrence of metastases.

Occurrence of metastases in patients[a] with:		
Monoclonal antibody	Positive prim. tumours	Negative prim. tumours
M-2-2-4	12/52 (23%)	1/4 (25%)
M-2-7-6	2/10 (20%)	7/23 (30%)
M-2-10-15	2/11 (18%)	7/27 (26%)
A-10-33	7/10 (70%)	6/33 (18%)
A-1-43	10/19 (53%)	5/37 (14%)
HLA DR[b]	13/29 (45%)	4/31 (8%)

[a] Observation period from 3–27 months
[b] 910D7 and OKIa1

Table VI. Changes in antigenic expression of melanoma lines after treatment with biological response modifiers

Monoclonal antibody	Phorbol myristine acetate	Retinoic acid	Dimethyl-formamide
M-2-2-4	0	0	0
M-2-7-6	+	–	0
M-2-10-15	+	–	+
A-10-33	+	0	0
A-1-43	0	0	–

0 = no change; + = upregulation; – = downregulation

Antibodies Generated Against Membrane Extracts of Melanoma Tissues

As it became evident that the monoclonal antibodies raised against melanoma cell lines reacted only in limited numbers with melanoma biopsies, and that many displayed a broad cross-reactivity with normal tissue, another set of monoclonal antibodies was raised against membrane preparations of various melanoma biopsies. So far we have been unable to detect melanoma-specific antigens, but we found antigens that fell into one of the previously defined categories. In addition, we have obtained antibodies that define new specificities of nevocellular antigens (table VII). The two antibodies generated against melanoma tissue show a different reaction profile on tissues as the two antibodies raised against cell lines.

Another example is given by the antibodies against endothelial antigens. The two antibodies generated against melanoma tissue display broader cross-reactivity than the cell line antibody (table VIII).

Table VII. Comparison of monoclonal antibodies directed against antigens of the *nevocellular* group

	Primary melanoma	Metastatic	Nevi
K-1-2-58[b]	50%[a]	38%	38%
K-3-8-7[b]	41%	11%	50%
M-2-2-4[c]	92%	91%	100%
M-2-7-6[c]	30%	28%	37%

[a] % cases positive
[b] Generated against melanoma biopsies
[c] Generated against melanoma cell lines

Table VIII. Comparison of monoclonal antibodies directed against antigens of the *endothelial* group

Monoclonal antibody	Primary melanoma	Metastatic	Nevi
A-10-33-1[a]	22%	21%	0%
H-2-5-47[b]	25%	47%	0%
H-2-7-33[b]	40%	58%	0%

[a] Generated against melanoma lines
[b] Generated against melanoma biopsy

Conclusions

From our experience with the production and specificity analysis of monoclonal antibodies against human melanoma cells we can draw the following conclusions:

(1) The monoclonal antibodies raised against melanoma cell lines detect no melanoma-specific antigens, but rather differentiation antigens that are also expressed by certain normal cells or other tumour cells.

(2) The *nevocellular I* antigen comes closest to the definition of a lineage-specific melanoma antigen. The *nevocellular II* and the *neural* antigens can also be classified as lineage antigens of neuroectodermal origin and have a more restricted occurrence.

(3) The *endothelial*, the *basal cell* and also the *HLA-DR* antigens are expressed by normal cells other than the neural crest, but are expressed by melanoma cells in the course of tumour progression.

(4) The diagnostic and prognostic value of certain antibodies is revealed only by extensive studies on melanoma biopsies of different stages: thus it turned out that the *endothelial*, the *basal cell* and the *HLA-DR* antigen are significantly more expressed in high risk melanomas. The *basal cell* antigen is preferentially expressed in locoregional metastases.

(5) The melanoma-associated antigens with restricted expression in vivo can be modulated in vitro by biological response modifiers. This shows the plasticity of tumour phenotypes which can be modulated by many environmental signals.

(6) By immunization with melanoma biopsies different specificities

of monoclonal antibodies were obtained as compared to immunization with established cell lines.

For the description of the complex cellular and molecular events of tumour progression one antigen or one monoclonal antibody is not sufficient. It cannot yet be decided whether we need 20, 30 or more different antibodies to cover the whole spectrum of tumour phenotypes.

Summary

The expression of melanoma-associated antigens in primary and metastatic melanomas was studied immunohistologically with a panel of monoclonal antibodies raised against melanoma cell lines and melanoma biopsies. The monoclonal antibodies were classified according to their cross-reaction patterns in normal tissues. A nevocellular antigen, recognized by the monoclonal antibody M-2-2-4, was broadly expressed by the majority of melanomas of different stages, whereas the staining with other monoclonal antibodies disclosed a heterogeneity within individual tumours and among the melanomas tested. Regarding clinical parameters, we found a relationship between the expression of some melanoma-associated antigens and the metastatic potential of primary melanomas. Comparing different metastases, a decrease of several melanoma-associated antigens was observed in advanced stages of disease.

The expression of melanoma-associated antigens which are differently expressed *in situ* could be modulated *in vitro* when melanoma cells were treated with biological response modifiers (phorbol myristine acetate, retinoic acid, dimethylformamide).

References

1 Seibert, E.; Sorg, C.; Happle, R.; Macher, E.: Membrane associated antigens of human malignant melanoma III. Specificity of human sera reacting with cultured melanoma cells. Int. J. Cancer *19:* 172–178 (1977).
2 Sorg, C.; Brüggen, J.; Seibert, E., Macher, E.: Membrane associated antigens of human malignant melanoma. IV. Changes in expression of antigens on cultured melanoma cells. Cancer Immunol. Immunother. *3:* 259–271 (1978).
3 Brüggen, J.; Sorg, C.; Macher, E.: Membrane associated antigens of human malignant melanoma V. Serological typing of cell lines using antisera from nonhuman primates. Cancer Immunol. Immunother. *5:* 53–62 (1978).
4 Brüggen, J.; Macher, E.; Sorg, C.: Expression of surface antigen and its relation to parameters of malignancy in human malignant melanoma. Cancer Immunol. Immunother. *10:* 121–127 (1981).
5 Sorg, C.; Brüggen, J.; Terbrack, D.; Vakilzadeh, F.; Suter, L.; Macher, E.: Cell surface structure and state of malignancy in human malignant melanoma; in Reisfeld, Ferrone, Melanoma antigens and antibodies, pp. 339–354 (Plenum Press, New York 1982).

6 Stenzinger, W.; Brüggen, J.; Macher, E.; Sorg, C.: Tumour angiogenic factor (TAF) production in vitro and growth in the nude mouse by human malignant melanoma. Eur. J. Cancer clin. Oncol. *19:*649–656 (1983).
7 Suter, L.; Brüggen, J.; Sorg, C.: Use of an enzyme-linked immunosorbent assay (ELISA) for screening of hybridoma antibodies against cell surface antigens. J. immunol. Methods *39:*407–411 (1980).
8 Brüggen, J.; Sorg, C.: Detection of phenotypic differences on human malignant melanoma lines and their variant sublines with monoclonal antibodies. Cancer Immunol. Immunother. (in press, 1983).
9 Suter, L.; Bröcker, E.-B.; Brüggen, J.; Ruiter, D. J.; Sorg, C.: Heterogeneity of primary and metastatic human malignant melanoma as detected with monoclonal antibodies in cryostat sections of biopsies. Cancer Immunol. Immunother. (in press, 1983).
10 Bröcker, E. B.; Suter, L.; Sorg, C.: HLA-DR. expression in primary melanomas of the skin (submitted for publication).
11 Bröcker, E. B.; Suter, L.; Brüggen, J.; Vakilzadeh, F.; Macher, E; Sorg, C.: Antigenic properties of low-risk and high-risk malignant melanomas (in preparation).
12 Brüggen, J.; Redmann, K.; Sorg, C.: Changes of biochemical and antigenic properties in human melanoma cells after treatment with biological response modifiers. J. Cancer Res. clin. Oncol. *105:*39 (1983).

Dr. Josef Brüggen, Abt. f. Experimentelle Dermatologie, Universitäts-Hautklinik, Von-Esmarch-Str. 56, D-4400 Münster (FRG)

Uses of Conventional and Monoclonal Antibodies to Intermediate Filament Proteins in the Diagnosis of Human Tumours

M. Osborn[1], E. Debus[1], M. Altmannsberger[2], K. Weber[1]

[1] Max Planck Institute for Biophysical Chemistry, Göttingen, FRG
[2] Department of Pathology, University of Göttingen, Göttingen, FRG

The purpose of this article is to outline a strategy for typing human tumours which has emerged out of basic biological research on the cytoskeleton. This strategy is based on subclassifying the so-called intermediate filaments (IFs), one of the cytoskeletal elements present in almost all cells of vertebrate origin. IFs can be distinguished from the other two fibrous elements of the cytoskeleton – microtubules and microfilaments – by their diameters in electron micrographs [20], and also by the use of antibodies in immunofluorescence microscopy [27]. Figure 1 shows intermediate filaments in a cell in culture – note the abundance of the filaments and how these filaments can "fill" the cytoplasm. Intermediate filaments are also present in cells in tissues and again can readily be distinguished by immunofluorescence microscopy or other immunocytological methods. Biochemical methods also show that IFs are major cellular proteins. Immunological, biochemical and subsequent protein sequence studies started five years ago have clearly established that cells in tissues can be subdivided into six major subclasses due to their IF content. These are shown in table I. Thus epithelial tissues – whether or not they keratinize – are distinguished by (cyto)keratins; most but not all neurones contain neurofilaments; astrocytes and a few other cells of glial origin contain glial fibrillary acidic protein; myogenic derivatives can be distinguished by the presence of desmin (but note that some vascular smooth muscle cells do not have desmin); "mesenchymal" derivatives as well as a few other non-epithelial cell types contain vimentin. In addition there is a sixth class of cells that includes, for instance, cells of the inner cellular mass of early embryo and some neurones, where IFs cannot be found. The names (cyto)keratin, neurofilament, GFA, desmin, and vimentin are thus the trivial names used to describe the major proteins which constitute the intermediate filaments present in the different cell types. Table I also lists the molecular weights of these proteins and shows that two of the classes contain multiple polypeptides. Thus in the human 19 different cytokeratins have been reported [24], and three neurofilament proteins have been described. The striking feature about table I is that the subdivisions follow histological principles, i.e., they closely approximate the major subdivisions made in histological textbooks. That IFs should influence the structural appearance of the cytoplasm is not unexpected given their abundance and their insolubility. That the subdivisions, however, should agree so well with those used in pathological diagnosis was both a surprise and a delight to those laboratories involved in this sort of analysis since it immediately suggested that IFs might be a good marker of histogenetic origin both for abnormal tissues and for malignant disease.

Before describing such applications it should be stressed that in the last five years both immunological and proteinchemical studies have shown that the intermediate filament proteins form a multigene

Table I. Types of intermediate filaments

Type	Protein	Examples in situ
1. Epithelial	(Cyto) keratin multiple polypeptides 45–60K	Keratinizing and nonkeratinizing epithelia
2. Neuronal	Neurofilament 68 K, 145 K and 200 K	Most but probably not all neurones in central and peripheral nerves
3. Glial	Glial fibrillary acidic protein 55K	Astrocytes, Bergmann glia (some also V^+)
4. Muscle	Desmin 53K	Sarcomeric muscle; smooth muscle; vascular smooth muscle cells can be D^+V^+, D^+V^- or D^-V^+
5. Mesenchymal	Vimentin 57K	Fibroblasts, chondrocytes, macrophages, endothelial cells etc.
6. No IFs		Cells of the early embryo, certain neurones

V: vimentin; D: desmin

family of related but distinct polypeptides. Thus, even though by looking at an electron micrograph in which IFs are visible, it is impossible to say to which type they belong, they can be assigned to one or other of the subclasses given in table I by the appropriate immunological, biochemical or proteinchemical classification. Proteinchemical studies in particular have established that these IF proteins share related sequences [15, 16, 32] and also conform to a common structural principle, first illustrated for desmin [15]. Thus desmin, vimentin, GFA, the three neurofilament proteins, some cytokeratins, and the two prototypes of hard α-keratin, the wool components 7 c and 8 c-1, contain a long rod-like domain some 38K in length whose structure is dictated by α-helical arrays able to form coiled-coils [15, 16, 32]. It is this part of the polypeptide which is responsible both for the α-type X-ray diffraction pattern displayed by IFs and for other shared properties such as their common morphologies, the 20 nm lateral periodicity seen after metal shadowing and their ability to reassemble in vitro. Homologies in the rod-like region are in the range of 50 to 60% for the non-epithelial IF proteins, i.e. desmin, vimentin, GFA and the NF68K polypeptide, whereas epithelial IF proteins, i.e., the cytokeratins and the wool components show a lower degree of homology. In contrast to the rod domain, the aminoterminal headpiece and the carboxyterminal tailpiece are hypervariable in sequence and in length when different IF proteins are compared. These regions of the molecule are not in an α-helical conformation, and the tailpiece in particular shows an extreme variation in length (from approximately 5,500 u in desmin to approximately 150,000 u in the neurofilament 200,000 u protein [15, 16, 32]). The protein sequence studies of IF proteins, as well as recent DNA sequence studies (e. g. [18]) confirm the differences between different IF proteins but also point out the necessity in immunolgical studies of using only highly specific antibodies that clearly react with only one IF type in a broad cross-species-specific manner. Such antibodies have been described from a number of laboratories [12, 19, 28], and recently we and others have started to apply monoclonal technology so as to isolate an equivalent set of monoclonals. One way to check that antibodies are indeed specific is to characterize them on tissues containing more than one cell type, and to show that each antibody stains only the expected cell type (see e.g. pictures of human large bowel and rat tongue; refs. [2, 28] respectively).

Results in which the IFs present in many cell types in adult and

embryonic animals, and in humans have been typed, have been summarized elsewhere [12, 19, 28]. In general, cells in the adult in situ contain only one IF type with the subdivisions corresponding to those shown in table I. Currently there are three major exceptions to this general rule:
(1) Astrocytes which may contain GFA and vimentin,
(2) certain vascular smooth muscle cells which contain desmin and vimentin,
(3) horizontal cells of the retina which contain NFs and vimentin.

Other exceptions are apparent during development. Of particular interest are the lack of IFs in early embryo until (cyto)keratin filaments are first observed in the trophoectoderm of the blastocyst [12, 28]; the finding that cells of parietal endoderm may have both cytokeratin and vimertin [22], and the finding of NFs and vimentin during embryonic development in the chicken [19]. Using such methods one can obviously begin to construct cell lineages, and can also characterize and identify particular cell types in complex tissues such as brain, or retina [32]. IF typing is also of use in characterizing cells present in primary or secondary cell cultures.

What happens with pathological material and in particular with tumours? Does one keep the IF typical of the cell of origin? Or does the tumour acquire additional IF types, either in the primary or in metastases? To answer such questions a collaboration was started between our group at the Max Planck Institute, and pathologists at the University of Göttingen. Our initial problem was to find an easy way to preserve tumour tissue, and usually we used either frozen sections or a method in which the tumour sample is fixed for 24 h in alcohol and then embedded in paraffin [2]. The IF antibodies used initially were affinity-purified on their original antigens; more recently we have begun to use monoclonal antibodies directly from the hybridoma supernatant fluids. The actual procedure, once sections have been cut, can be performed in less than 1.5 h. Our initial study confirmed that the antibodies define the tissue specificity also on human material [2]; an essential point, since some species-specific amino acid exchanges clearly exist for IF proteins. Examination of a larger series of both breast [1] and gastrointestinal [4] tumours established that broad specificity cytokeratin antibodies were excellent markers for carcinomas. Excluding two carcinomas that were improperly fixed, tumour cells in 18/18 (100%) of the breast carcinomas were cytokeratin-positive, and tumours cells in 25/25 (100%) of the gastrointestinal carcinomas were cytokeratin-positive. Interestingly signet ring cell carcinomas of the stomach could be particularly easily identified using such techniques [4]. Vimentin did not stain any of the tumour cells in the carcinomas that were examined. Vimentin did, however, give a very strong reaction on tumour cells of non-muscle sarcomas [3]; thus tumours such as lymphoma, liposarcoma, tumourous leukemia, and dermatofibrosarcoma protuberans are vimentin-positive, but negative for other IF types. *Gabbiani et al.* [13] have also shown the use of keratin and vimentin antibodies to discriminate carcinoma from non-muscle sarcoma. Of particular interest has been the use of desmin antibodies to diagnose rhabdomyosarcoma. We initially reported six cases of rhabdomyosarcoma that were all desmin-positive [3]. Currently, we have examined some 29 specimens sent to us by clinicians as possible rhabdomyosarcomas. Of these, 16 specimens were desmin-positive. Taking this finding together with other pathological criteria, these tumours have been classified as rhabdomyosarcoma. The remaining 13 specimens were desmin-negative, and because of this finding, together with other pathological criteria, are thought not to be rhabdomyosarcoma. Thus far our results support the use of desmin as a marker for rhabdomyosarcoma, and suggest that it is currently the best one available for this tumour. NF antibodies specific for one or more NF polypeptides clearly identify some tumours originating from sympathetic neurones, e.g. pheochromocytoma, ganglioneuroblastoma and some neuroblastomas [26, 29]. Whether all neuroblastomas are neurofilament-positive, or whether some may lack NFs, remains to be determined. If one reviews the use of IF antibodies on human tumours several points seem important. First, broadly similar results are now available from a number of laboratories using independently prepared antibodies for immunohistological typing of tumours ([1–8, 23, 26, 30]; for review see [29]). Second, a few antibodies have been used in larger clinical trials (e.g. [25]), again with very promising results. Third, so far at least for primary tumours and for solid metastases it seems as though the IF type (or types) typical of the cell of origin is retained (above references). In these cases additional IF types seem not to be acquired (note, however, ref. [30] for tumours growing as ascites).

Are antibodies to IFs of use in routine diagnostic pathology? Some 95% of tumours that come across the pathologist's desk can be

diagnosed without difficulty using conventional histological stains. What of the other 5%? Here immunological diagnosis can be of value if the distinctions that can be made using the antibodies lead to a difference in treatment for the patient. Obvious examples where IF typing could lead to improved diagnosis include
(1) undifferentiated carcinoma (keratin-positive) vs. lymphoma (vimentin-positive),
(2) the small cell tumours of childhood. Here a distinction is possible between rhabdomyosarcoma (desmin-positive) vs. lymphoma or Ewing's sarcoma (both vimentin-positive) vs. neuroblastoma (NF-positive; some cases perhaps lack all IF types) vs. Wilms tumour (blastema cells in most cases have both cytokeratin and vimentin),
(3) tumours such as signet ring carcinoma of the stomach where only very few tumour cells may be present,
(4) cytological specimens.

Other examples can be found in some of the references. Interestingly IF typing has already led to a revision or reconsideration of the original light microscopic diagnosis in some cases, and the number of such cases can be expected to increase.

In addition, IF typing of certain tumours, as well as of normal cells in tissues has indicated that at least certain tumours arise by the selective multiplication of a particular identifiable cell type in the normal tissue.

In certain instances monoclonal antibodies can offer advantages over their conventional counterparts. Thus, in principle they provide a readily exchangeable reagent of unlimited quantity where the epitope recognized by the antibody has to be defined only once. In addition, at least for those IF classes which contain more than one polypeptide, i.e. the neurofilaments and the cytokeratins, further subdivision of these classes may be easier with monoclonal rather than with conventionally prepared antibodies. For these reasons we started in 1981 to develop a set of monoclonal antibodies, each member of which would recognize one and only one IF type. So far we have isolated and characterized monoclonal antibodies specific for each of the three neurofilament polypeptides, desmin, GFA, and vimentin [9, 10, 10a, 10b]. These mouse monoclonal antibodies behave like their conventional counterparts and can also be used to type cells in both normal tissues and in tumours with similar results. Thus, for instance, the monoclonal desmin antibodies identify the tumour cells in rhabdomyosarcoma (figure 2 a), the monoclonal vimentin antibody identifies tumour cells in infantile fibrosarcoma (figure 2 b), the monoclonal 200K neurofilament antibody [9] identifies the tumour cells present in tumours derived from sympathetic neurones (figure 2 c; [26]). In addition we have also isolated and characterized cytokeratin antibodies CK1 – CK4 [10] which specifically recognize a single cytokeratin polypeptide (no. 18 in the nomenclature of ref. [24]). In agreement with the demonstration that this cytokeratin is not present in stratified squamous epithelia, the CK1 – CK4 antibodies identify simple epithelia, as well as transitional epithelia of the bladder, but do not recognize stratified squamous epithelia from skin or oesophagus. Recently CK1 – CK4 have been used to subclassify carcinomas [11]. Thus tumours originating from simple epithelia are strongly stained by CK1 – CK4 (e.g. the ductal carcinoma of the breast shown in figure 2d). Tumours such as squamous cell carcinoma of the skin, tongue or oesophagus derived from stratified squamous epithelia are not stained, while in two of the cases so far examined, i.e. squamous

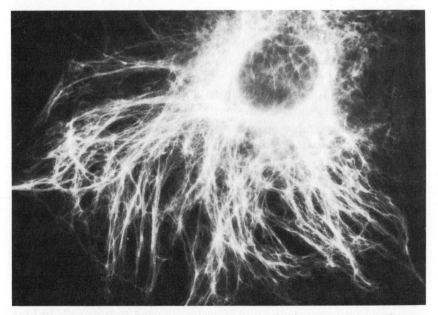

Fig. 1. Intermediate filaments in a fibroblast growing in culture. Immunofluorescence with antibody to vimentin. × 780.

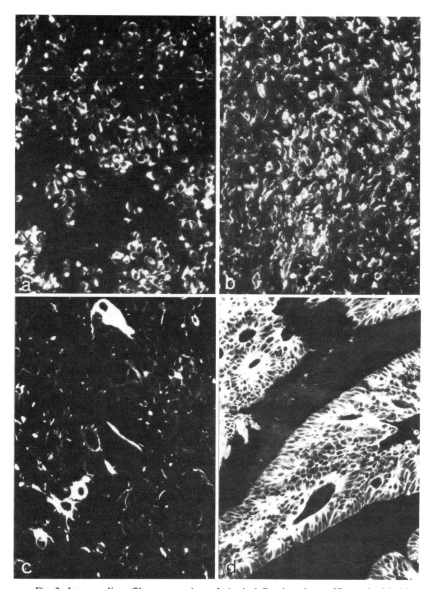

Fig. 2. Intermediate filament typing of alcohol-fixed and paraffin-embedded human tumour material with monoclonal antibodies. *(a)* rhabdomyosarcoma stained with monoclonal antibody specific for desmin; × 240; *(b)* infantile fibrosarcoma stained with monoclonal antibody specific for vimentin; × 240; *(c)* pheochromocytoma stained with monoclonal antibody specific for the neurofilament 200K polypeptide; × 240; *(d)* ductal carcinoma of the breast stained with monoclonal antibody specific for cytokeratin 18; × 150.

Table II. Typing of human tumours using antibodies to intermediate filaments, showing selected examples from tumours typed for IF content in Göttingen. IF typing clearly distinguishes the major human tumour groups, i. e. carcinomas are cytokeratin-positive, non-muscle sarcomas are vimentin-positive, muscle sarcomas desmin-positive, glial-derived tumours GFA-positive, and tumours such as pheochromocytoma, ganglioneuroblastoma and at least some neuroblastomas are neurofilament-positive. For further details see refs. [1–4, 26, 29]. For similar results with cytological specimens see ref. [5]. Results in this table were obtained using conventional antibodies specific for each IF tpye; for similar results with monoclonal antibodies see figure 2

Tumour	No. of cases	Cytokeratin	Vimentin	Desmin	GFA	NFs
Breast carcinoma	20	+	–	–	–	–
Stomach carcinoma	10	+	–	–	–	–
Large bowel carcinoma	20	+	–	–	–	–
Angiosarcoma	1	–	+	–	–	–
Dermatofibrosarcoma protuberans	1	–	+	–	–	–
Malignant fibrous histiocytoma	4	–	+	–	–	–
Pleomorphic liposarcoma	1	–	+	–	–	–
Malignant Schwannoma	1	–	+	–	–	–
Synovial sarcoma (monophasic)	3	–	+	–	–	–
Fibrosarcoma	2	–	+	–	–	–
Ewings sarcoma	1	–	+	–	–	–
Malignant lymphoma	5	–	+	–	–	–
Rhabdomyosarcoma	16	–	(+)	+	–	–
Ependymoblastoma	1	–	–	–	+	–
Pheochromocytoma	5	–	–	–	–	+
Ganglioneuroblastoma	3	–	–	–	–	+
Neuroblastoma	9	–	–	–	–	+(6)–(3)

+ all cases positive; – negative; (+) some cases show positive staining

cell carcinoma of the epiglottis and of the cervix uteri, the least differentiated tumour cells are stained, whereas the most differentiated are not. This study, as well as studies using monoclonals to cytokeratins in other laboratories [4, 17, 21, 22, 31] thus open the way to a further subdivision of carcinomas using cytokeratin typing. Ideally if it proves possible to isolate cytokeratin monoclonals specific for certain tissues, it may be possible to decide from a lymph node metastasis not only that it is derived from a primary carcinoma, but also to pinpoint the site.

Summary

Tumours retain the IF type typical of the cell of origin. Thus typing of the IFs present in tumour cells can yield information about the histogenetic origin. IF typing can distinguish the major human tumour groups. It can in addition provide an unambiguous and rapid characterization in cases such as the small cell tumours of childhood, and can discriminate between undifferentiated carcinoma and lymphoma. Equivalent results are obtained with either conventionally prepared antibodies or with their monoclonal counterparts. Monoclonal antibodies, however, may allow additional subdivision to be made, particularly of carcinomas.

Acknowledgments: We thank *Sabine Schiller* and *Susanne Isenberg* for technical help, and Professor *A. Schauer* for discussion. This work has been supported by the Max-Planck-Gesellschaft and the Deutsche Forschungsgemeinschaft.

References

1. Altmannsberger, M.; Osborn, M.; Hölscher, A.; Schauer, A.; Weber, K.: The distribution of keratin type intermediate filaments in human breast cancer: an immunohistological study. Virchows Arch. Abt. B Zellpath. *37:* 277–284 (1981).
2. Altmannsberger, M.; Osborn, M.; Schauer, A.; Weber, K.: Antibodies to different intermediate filament proteins: cell type-specific markers on paraffin-embedded human tissues. Lab. Invest. *45:* 427–434 (1981).
3. Altmannsberger, M.; Osborn, M.; Treuner, J.; Hölscher, A.; Weber, K.; Schauer, A.: Diagnosis of human childhood rhabdomyosarcoma by antibodies to desmin the structural protein of muscle specific intermediate filaments. Virchows Arch. Abt. B. Zellpath. *39:* 203–215 (1982).
4. Altmannsberger, M.; Weber, K.; Hölscher, A.; Schauer, A.; Osborn, M.: Antibodies to intermediate filaments as diagnostic tools: human gastrointestinal carcinomas express keratin. Lab. Invest. *46:* 520–526 (1982).
5. Altmannsberger, M.; Osborn, M.; Droes, M.; Weber, K.; Schauer, A.: Diagnostic value of intermediate filament antibodies in clinical cytology. Klin. Wschr. (in press, 1983).
6. Battifora, H.; Sun, T.-T.; Bahu, R.; Rao, S.: The use of anti-keratin antiserum in tumour diagnosis. Lab. Invest. *42:* 100–101 (1981).
7. Bignami, A.; Dahl, D.; Rueger, D. C.: Glial fibrillary acidic protein (GFA) in normal neural cells and in pathological conditions. Adv. Cell. Neurobiol. *1:* 285 (1980).
8. Caselitz, J.; Jänner, M.; Breitbart, E.; Weber, K.; Osborn, M.: Malignant melanomas contain only the vimentin type of intermediate filament: implications for histogenesis and diagnosis. Virchows Arch. Abt. A Path. Anat. *400:* 43–51 (1983).
9. Debus, E.; Flügge, G.; Weber, K.; Osborn, M.: A monoclonal antibody specific for the 200K polypeptide of the neurofilament triplet. EMBO J. *1:* 41–45 (1982).
10. Debus, E.; Weber, K.; Osborn, M.: Monoclonal cytokeratin antibodies that distinguish simple from stratified human epithelia: characterization on human tissues. EMBO J. *1:* 1641–1647 (1982).

10a Debus, E.; Weber, K.; Osborn, M.: Monoclonal antibodies to desmin, the muscle-specific intermediate filament protein. EMBO J. *2:* (in press, 1983).
10b Debus, E.; Weber, K.; Osborn, M.: Monoclonal antibodies specific for glial fibrillary acidic (GFA) protein and for each of the neurofilament triplet polypeptides. Differentiation (in press, 1983).
11 Debus, E.; Moll, R.; Franke, W. W.; Weber, K.; Osborn, M.: Immunohistochemical distinction of human carcinomas by cytokeratin typing with monoclonal antibodies. Am. J. Path. (in press, 1983).
12 Franke, W. W.; Schmid, E.; Schiller, D. L.; Winter, S.; Jarasch, E.; Moll, R.; Denk, H.; Jackson, B. W.; Illmensee, K.: Differentiation patterns of expression of proteins of intermediate-sized filaments in tissues and cultured cells. Cold Spring Harbor Symp. Quant. Biol. *46:* 413–430 (1982).
13 Gabbiani, G.; Kapanci, Y.; Barazzone, P.; Franke, W. W.: Immunochemical identification of intermediate-sized filaments in human neoplastic cells. Am. J. Path. *104:* 206–216 (1981).
14 Gatter, K. C.; Abdulaziz, Z.; Beverley, P.; Corvalan, J. R. F.; Ford, C.; Lane, E. B.; Mota, M.; Nash, J. R. G.; Pulford, K.; Stein, H.; Taylor-Papadimitriou, J.; Woodhouse, C.; Mason, D. Y.: Use of monoclonal antibodies for the histopathological diagnosis of human malignancy. J. clin. Path. *35:* 1253–1267 (1982).
15 Geisler, N.; Kaufmann, E.; Weber, K.: Proteinchemical characterization of three structurally distinct domains along the protofilament unit of desmin 10 nm filaments. Cell *30:* 277–286 (1982).
16 Geisler, N.; Weber, K.: The amino acid sequence of chicken muscle desmin provides a common structural model for intermediate filament proteins including the wool α-keratins. EMBO J. *1:* 1649–1656 (1982).
17 Gigi, O.; Geiger, B.; Eshhar, Z.; Moll, R.; Schmid, E.; Winter, S.; Schiller, D. L.; Franke, W. W.: Detection of a cytokeratin determinant common to diverse epithelial cells by a broadly cross reacting monoclonal antibody. EMBO J. *1:* 1429–1437 (1982).
18 Hanukoglu, I.; Fuchs, E.: The cDNA sequence of a human epidermal keratin: divergence of sequence but conservation of structure among intermediate filaments. Cell *31:* 243–252 (1982).
19 Holtzer, H.; Bennett, G. S.; Tapscott, S. J.; Croop, J. M.; Toyama, Y.: Intermediate-sized filaments: changes in synthesis and distribution in cells in myogenic and non-myogenic lineages. Cold Spring Harbor Symp. Quant. Biol. *46:* 317–330 (1982).
20 Ishikawa, H.; Bischoff, R.; Holtzer, H.: Mitosis and intermediate-sized filaments in developing skeletal cells. J. Cell Biol. *38:* 538–555 (1968).
21 Lane, E. B.: Monoclonal antibodies provide specific intramolecular markers for the study of epithelial tonofilament organization. J. Cell Biol. *92:* 665–673 (1982).
22 Lane, E. B.; Hogan, B. L. M.; Kurkinen, M.; Garrels, J. I.: Coexpression of vimentin and cytokeratins in parietal endoderm cells of the early mouse embryo. Nature *303:* 701–704 (1983).
23 Miettinen, M.; Lehto, V.-P.; Badley, R. A.; Virtanen, I.: Expression of intermediate filaments in soft tissue sarcomas. Int. J. Cancer *29:* 541–546 (1982).
24 Moll, R.; Franke, W. W.; Schiller, D. L.; Geiger, B.; Krepler, R.: The catalogue of human cytokeratin polypeptides: patterns of expression of cytokeratins in normal epithelia, tumors and cultured cells. Cell *31:* 11–24 (1982).
25 Nagle, R. B.; McDaniel, K. M.; Clark, V. A.; Payne, C. M.: The use of antikeratin antibodies in the diagnosis of human neoplasms. Am. J. clin. Path. *79:* 458–466 (1983).

26　Osborn, M.; Altmannsberger, M.; Shaw, G.; Schauer, A.; Weber, K.: Various sympathetic derived human tumours differ in neurofilament expression. Virchows Arch. Abt. B Zellpath. *40:* 141–156 (1982).
27　Osborn, M.; Franke, W. W.; Weber, K.: The visualization of a system of 7–10 nm thick filaments in cultured cells of an epitheloidal line (PtK2) by immunofluorescence microscopy. Proc. natn. Acad. Sci. USA *74:* 2490–2494 (1977).
28　Osborn, M.; Geisler, N.; Shaw, G.; Sharp, G.; Weber, K.: Intermediate filaments. Cold Spring Harbor Symp. Quant. Biol. *46:* 413–429 (1982).
29　Osborn, M.; Weber, K.: Tumour diagnosis by intermediate filament typing. Lab. Invest. *48:* 372–394 (1983).
30　Ramaekers, F. C. S.; Haag, D.; Kant, A.; Moesker, O.; Jap, P. H. K.; Vooijs, G. P.: Co-expression of keratin and vimentin-type intermediate filaments in human metastatic carcinoma cells. Proc. natn. Acad. Sci. USA *80:* 2618–2622 (1983).
31　Tseng, S. C. G.; Jarvinen, M. J.; Nelson, W. G.; Huang, J. W., Woodcock-Mitchell, J.; Sun, T. T.: Correlation of specific keratins with different types of epithelial differentiation: monoclonal antibody studies. Cell *30:* 361 (1982).
32　Weber, K.; Shaw, G.; Osborn, M.; Debus, E.; Geisler, N.: Neurofilaments, a subclass of intermediate filaments: structure and expression. Cold Spring Harbor Symp. Quant. Biol. *48:* (in press, 1983).

Mary Osborn, Ph. D., Max-Planck-Institut für Biophysische Chemie, D-3400 Göttingen (FRG)

Immunodiagnosis of Human Solid Tumours

M. Herlyn, M. Blaszczyk, H. Koprowski

The Wistar Institute of Anatomy and Biology, Philadelphia, Pa., USA

Several distinct applications of immunologic procedures for the diagnosis of human tumours might be considered. First, some immunodiagnostic approaches might be useful in the early detection of cancer cases by screening high-risk population groups. Second, immunodiagnostic procedures may aid in distinguishing between patients with cancer and those with benign diseases. Third, immunoassays might be used to localize a tumour, to determine prognosis and type of therapy, to monitor the patients' response to therapy and to detect possible early recurrences [1].

Monoclonal antibodies have been developed in our laboratory and by others for the diagnosis of human solid tumours. These reagents can be used to localize tumours with nuclear imaging techniques [2], to visualize malignant cells in histological sections of lesions [3, 4], and to detect tumour-associated antigens in sera of patients with malignancies [5–7].

Certain antigens that have been identified using monoclonal antibodies in the sera of cancer patients are also released by cells of the same tumour type maintained in tissue culture. These shed antigens are expressed on the surface of in vitro grown cells as shown, for example, in figure 1, or they can be restricted to the cytoplasm of cells. Once the reactivity pattern of monoclonal antibodies on a variety of target preparations is determined, different radioimmunoassays for the detection of antigens circulating in patients' sera can be applied (table I). The antibody inhibition assay [6, 8] which is generally used in early screening studies, does not require purified antibody or antigen. However, this assay is suitable only for antibodies with high binding affinities and requires an optimal target preparation, involving costly cell extraction procedures, for each antibody. The double determinant immunoassay (DDIA) [9] has replaced the antibody, as well as the antigen, inhibition assay for most studies. The DDIA, which is shown in more detail in figure 2, requires only small quantities of crude ascitic fluid of mice bearing hybridoma tumours. The antibody is absorbed to a solid phase such as polystyrene beads ("antigen catcher"), incubated with serum, and a purified second antibody is added to "trace" the antigen. The same antibody can be used

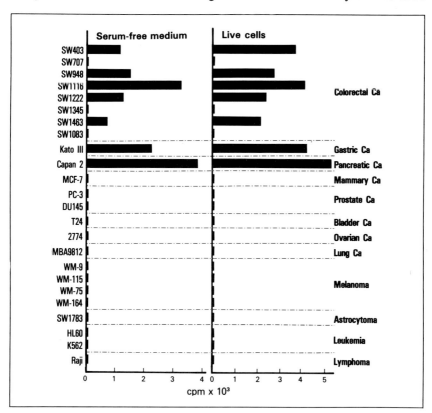

Fig. 1. Binding of monoclonal antibody 19-9 to live cells and serum-free medium. Live cells in suspension (2.5×10^5 cells/50 µl) were used as targets in indirect radioimmunoassay (RIA) (upper panel). Spent SSFM of cells (4 µg protein/50 µl) was dried and fixed to flexible microtiter plates and used as targets in solid-phase RIA (lower panel). Binding of monoclonal antibodies was detected using ^{125}I-labelled rabbit Ig anti-mouse F(ab')$_2$ (30,000–35,000 cpm). Results represent cpm minus control cpm (P3 × 63Ag8 supernatant).

Table I. Immunoassay protocols for the detection of tumour-associated antigens circulating in patients

Antibody inhibition assay	DDIA	Anti-idiotypic inhibition assay	Antigen inhibition assay
1. Mix MAb with serum	1. 1st MAb on solid-phase	1. MAb (idiotype) on solid-phase	1. MAb on solid-phase
2. Transfer to target	2. Add serum	2. Add serum	2. Add serum
3. Test for decrease of binding of MAb in indirect assay	3. Add labelled 2nd MAb	3. Add labelled anti-idiotype MAb	3. Add antigen labelled
	4. Test for binding of labelled antibody	4. Test for decrease of binding of anti-idiotype MAb	4. Test for decrease of binding of labelled antigen

as both "catcher" and "tracer" if the antigenic determinant detected by the monoclonal antibody is repeatedly present on the molecule, e.g., carbohydrates, or if the antigen circulates in the serum in an aggregated form. Tracing of antigen bound to the solid phase is optimal using an antibody that binds to a different determinant. If a second tracer is not available, an anti-idiotype inhibition assay can be used. However, this assay requires the production and purification of an antibody (monoclonal anti-idiotype) and, as such, is too costly for most approaches.

Immunization of mice with cells of gastrointestinal tumours and fusion of the immune splenocytes with mouse myeloma cells [10, 11] has generated monoclonal antibodies that, in addition to binding to gastrointestinal tumour cells, bind to blood group or blood group-related determinants (table II). The antigens detected by these antibodies are released by gastrointestinal tumour cells grown in vitro, but not by cells of any other tumour type. All of these antigens can be found circulating in patients' sera, though the serum antigens may differ from those on tumour cells defined by the same monoclonal antibodies. For example, the antigen detected by antibodies 19-9 and 52a has been characterized on tumour cells as a monosialoganglioside [12], whereas in serum of tumour patients, the antigen circulates as a glycomucin with a molecular weight of 2×10^6 u [*J. Magnani et al.*, manuscript in preparation].

The presence of 19-9- or 52a-defined antigenic determinants in sera of patients is highly associated with gastrointestinal malignancies (table III). An arbitrary "cut-off" between positive and negative sera was chosen at 12% binding inhibition in antibody inhibition assays [6]

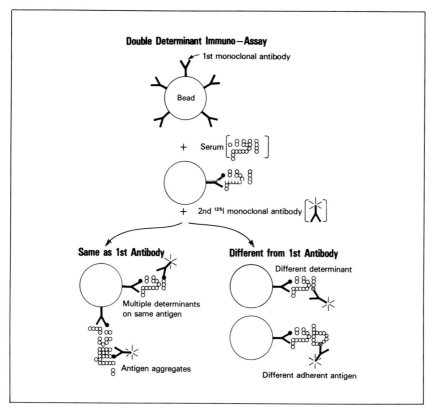

Fig. 2. DDIA for the detection of antigen(s) in patients' sera. The first antibody bound to the polystyrene bead is usually the 1:1,000 to 1:10,000 diluted ascitic fluid of a mouse bearing a hybridoma tumour. Sera are diluted 1:2 or higher. The purified second antibody is labelled with ^{125}I by the Iodogen method.

or 41 units in DDIA. Other investigators have obtained similar results [13]. Whereas the percentage of positivity was high in sera of advanced gastrointestinal malignancies, it was low (<20% positivity) in sera of patients with early recurrences or primary colorectal carcinoma.

To increase the sensitivity of detecting the 19-9-defined antigen, we developed several DDIAs in which antibodies against the blood group determinants listed in table II were used as catchers of antigen circulating in serum, and radiolabelled 19-9 antibody as antigen tracer. With the exception of the anti-H-antibodies, which were not tested, all anti-Lewis, anti-blood group and anti-lactofucopentaose

Table II. Monoclonal antibodies binding to carbohydrate determinants present on human cells and in patients' sera

Monoclonal antibodies	Antigen as defined on cell	Binding of antibodies to tumour cells grown in vitro
19-9 52a	Monosialoganglioside	Colorectal Ca, Gastric Ca, Pancreatic Ca
CO 29.3 CO 29.4 CO 30.1 CO 43.1	Lewis b	Colorectal Ca, Gastric Ca, Pancreatic Ca
CO 10 CO 43.2	Lewis b and H type I	Colorectal Ca, Gastric Ca, Pancreatic Ca
CO 51.4	Lewis a	Colorectal Ca, Gastric Ca, Pancreatic Ca
CO 51.2 CO 51.3	Lewis a and Lewis b	Colorectal Ca, Gastric Ca, Pancreatic Ca
33/25/1/17[a]	Blood group A	Colorectal Ca
PA 83-52 PA 23-48 PA 15-2 PA 66-18 GA 73.4	Blood group B	Colorectal Ca, Gastric Ca, Pancreatic Ca
GA 29-1 CO 56-22	Lactofucopentaose III (SSEA-1, Le^x)	Various adenocarcinomas, myelomonocytic cells
101[b]	H-type I	Not tested
102	H-type II	

[a] Obtained from C. Civin, Johns Hopkins Oncology Center, Baltimore, Md.
[b] Obtained from N. Richert, NIH, Bethesda, Md.

III antibodies catch an antigen in spent tissue culture medium or patients' sera that can also bind antibody 19-9. The number of false positive sera from healthy donors in these DDIAs ranges from 0% (anti-Lewis a and -Lewis b catcher /19-9 tracer) to 6% (anti-B catcher/19-9 tracer), whereas the percentage of positive sera from patients with advanced colorectal carcinoma ranges from 25% (anti-B) to 56% (anti-Lewis b). Moreover, the combination of anti-Lewis b antibody 10 and19-9 antibody increased the sensitivity of detecting early recurrences and primary colorectal carcinomas from 18% to 38% in sera of 55 patients.

In another series of experiments, we studied the sensitivity and specificity of monoclonal antibodies against carcinoembryonic antigen (CEA) and CEA-related antigens for the detection of circulating CEA in patients' sera. Fourteen monoclonal antibodies from 6 different laboratories were tested in different combinations in additive binding assays, and in only 2 of the 91 combinations did the antibodies apparently bind to the same or closely associated antigenic determinants (fig. 3). The antibodies could be divided into 6 groups, based on the molecular weights of antigen(s) immunoprecipitated from solubilized SW948 colorectal carcinoma cells (table IV). Antibody 39B6 of group II detects the "classical" CEA, whereas antibodies of groups III–VI immunoprecipitate additional lower molecular weight proteins. Antibodies that immunoprecipitate a 50K-u protein from tumour cells also characteristically bind to granulocytes. This 50K protein has been described as normal crossreacting antigen (NCA) [14].

DDIAs for the detection of CEA and CEA-related antigens in patients' sera indicate that antibodies of all groups are valuable reagents (group VI antibody was not tested). Antibodies of groups I and II apparently increase the specificity of detecting CEA by decreasing the percentage of false positive sera from patients with non-malignant diseases as compared to results with standard CEA assays using polyclonal antibodies. Antibodies of groups III, IV and V seem to increase the sensitivity of detecting malignancies, since up to 75% of sera from patients with advanced colorectal carcinoma showed elevated antigen levels using these antibodies, as compared with 57% with groups I and II antibodies. However, the percentage of false

Table III. Detection of a tumour-associated antigen by monoclonal antibodies against a monosialoganglioside

Patient's disease	No. of sera tested	% of sera with elevated antigen levels
Pancreatic Ca	55	89
Gastric Ca	17	70
Colorectal	285	64
Other malignancies	89	7.8
Non-malignant Gastrointestinal diseases	176	8.5
Healthy donors	237	2

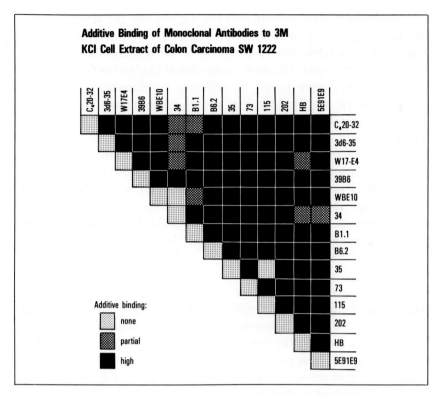

Fig. 3. Binding assay of monoclonal antibodies against CEA and related antigens. Two antibodies added simultaneously to a 3 M KCl extract of colon carcinoma cell line SW1222 showed no additive binding, indicating that they compete for the same or closely associated antigenic determinant on the molecule.

positives can be as high as 9% using groups III, IV and V antibodies, and 0% using antibodies of groups I or II.

We have recently produced monoclonal antibodies against colorectal, gastric and mammary carcinoma cells that do not bind to glycolipid antigens nor to CEA. Preliminary data suggest that the antigenic determinants detected by the 7 antibodies listed in table V are present in sera of patients with gastrointestinal tumours but not in sera of healthy donors.

Monoclonal antibodies that detect antigens in sera of patients with malignant melanoma have been obtained. Thus far, 4 antigenic classes have been detected in sera of patients (table VI). The high mo-

Table IV. Characteristics of CEA and related antigens which are present in sera of cancer patients

Group	Monoclonal antibodies		Molecular weight of protein(s) immunoprecipitated from solubilized SW948 colorectal carcinoma cells
	Example	Source	
I	3d635	The Wistar Institute	180K
II	39B6	J. Lingren, Helsinki	180K, 160K
III	B1.1	J. Schlom, NIH, Bethesda	180K, 160K, 50K
IV	$C_4$20-32	The Wistar Institute	180K, 160K, 50K 40K
V	B6.2	J. Schlom	50K
VI	B9F10	V. Zurawski, Malvern, Pa.	180K, and others

lecular weight proteoglycans are expressed and shed by most melanoma cells grown in vitro; these proteoglycans can be detected in patients with advanced melanoma. Elevated levels of HLA-DR were found in 16% of 31 sera from patients with advanced melanoma, but

Table V. Recently produced monoclonal antibodies that define antigens in sera of patients with gastrointestinal malignancies

Code no.	Monoclonal antibodies		Binding to in vitro grown cells of
	Isotype	Tumour cells used for immunization of mice	
CO 29.2	IgG1	Colorectal Ca	Colorectal Ca, Prostate Ca, Gastric Ca, Bladder Ca, Pancreatic Ca
CO 44.1	IgG2a	Colorectal Ca	Colorectal Ca, Gastric Ca, Ovarian Ca, Mammary Ca
CO 51.1	IgG1	Colorectal Ca	Colorectal Ca, Gastric Ca, Pancreatic Ca, Ovarian Ca, Mammary Ca
CO 14–72	IgG2a	Colorectal Ca	Colorectal Ca, Gastric Ca
CO 73.3	IgG2a	Gastric Ca	Most adenocarcinomas
BR 40.1	IgM	Breast Ca	Colorectal Ca, Mammary Ca, Granulocytes
CO 59.1	IgM	PEG-precipitate of serum from CRC-patient	

Table VI. Monoclonal antibodies against melanoma detect antigens in patients' sera

Antigen	Monoclonal antibodies	Assay used for detection in sera
Proteoglycan	ME 95–45	DDIA
	ME 31.3	
HLA-DR	13–17	Antibody inhibition
	37–7	DDIA
Gangliosides	ME 31.2	Indirect RIA
	ME 36.1	
Protein, 120K	ME 49.1	Antibody inhibition

in none of 11 sera from primary melanoma patients. Gangliosides of melanoma are shed in vitro by most melanoma cells but not by any other cells tested. The difficulty in detecting these antigens in sera results from the fact that shed gangliosides adhere to the surface of unrelated cells, such as erythrocytes and lymphocytes. p97, a well characterized melanoma-associated antigen, seems to be present at higher levels in sera of melanoma patients as compared to normal sera [15].

A 120 k u protein detected by antibody ME49.1 is shed in large quantities by melanoma cells and some carcinoma cells grown in vitro. Preliminary studies show that the antigen is present in normal sera but at much lower levels than in sera of melanoma patients.

Our studies indicate that monoclonal antibodies can be produced that detect a variety of antigens in sera of patients with malignancies. In some cases, such as pancreatic carcinoma, one antibody is sufficient to diagnose the malignancy; in other cases, such as colorectal carcinoma or melanoma, a panel of antibodies will be required as serodiagnostic reagents. Combinations of monoclonal antibodies with different specificities seem to increase the sensitivity of antigen detection, which would be most important for the early diagnosis of cancer.

Summary

Monoclonal antibodies have been produced that detect antigens circulating specifically in sera of patients with solid tumours such as colorectal carcinoma, gastric and pancreatic carcinoma as well as melanoma. Several double determinant immunoassays were developed that are sensitive, reproducible and allow the screening of large numbers of sera. For the serodiagnosis of certain tumours, such as pancreatic carcinoma, one antibody seems to be sufficient as a diagnostic reagent since 90% of patients' sera were positive. For other malignancies, such as colorectal carcinoma and melanoma, a panel of antibodies is necessary to detect specific antigen(s) in sera.

Acknowledgments: This work was supported in part by NIH grants CA-25874, CA-21124, and CA-10815. We thank our colleagues Drs. *D. Herlyn, Z. Steplewski,* and *A. Ross* for growth or purification of monoclonal antibodies. We also acknowledge the excellent technical assistance of *J. Bennicelli* and *M. Grosso.*

References

1. Herberman, R. B.: Immunologic tests in cancer. Cancer *68:* 688–698 (1977).
2. Herlyn, D.; Powe, J.; Alavi, A.; Mattis, J. A.; Herlyn, M.; Ernst, C.; Vaum, R.; Koprowski, H.; Radioimmunodetection of human tumor xenografts by monoclonal antibodies. Cancer Res. *43:* 2731–2735 (1983).
3. Atkinson, B. F.; Ernst, C. S.; Herlyn, M.; Sears, H. F.; Steplewski, Z.; Koprowski, H.: Immunoperoxidase localization of a monoclonal antibody-defined gastrointestinal monosialoganglioside. Cancer Res. *42:* 4820–4823 (1982).
4. Koprowski, H.; Brockhaus, M.; Blaszczyk, M.; Magnani, J.; Steplewski, Z.; Ginsburg, V.: Lewis blood-type may affect the incidence of gastrointestinal cancer. Lancet *ii:* 1332–1333 (1982).
5. Koprowski, H.; Herlyn, M.; Steplewski, Z.; Sears, H. F.: Specific antigen in serum of patients with colon carcinoma. Science *212:* 53–55 (1981).
6. Herlyn, M.; Sears, H. F.; Steplewski, Z.; Koprowski, H.: Monoclonal antibody detection of a circulating tumor-associated antigen. I. Presence of antigen in sera of patients with colorectal, gastric, and pancreatic carcinoma. J. clin. Immunol. *2:* 135–140 (1982).
7. Sears, H. F.; Herlyn, M.; del Villano, B.; Steplewski, Z.; Koprowski, H.: Monoclonal antibody detection of a circulating tumor-associated antigen. II. A longitudinal evaluation of patients with colorectal cancer. J. clin. Immunol. *2:* 141–149 (1982).
8. Steplewski, Z.; Chang, T. H.; Herlyn, M.; Koprowski, H.: Release of monoclonal antibody-defined antigens by human colorectal carcinoma and melanoma cells. Cancer Res. *41:* 2723–2727 (1981).
9. Brown, J. P.; Woodbury, R. G.; Hart, C. E.; Hellstrom, I.; Hellstrom, K. E.: Quantitative analysis of melanoma-associated antigen in p97 in normal and neoplastic tissues. Proc. natn. Acad. Sci. USA *78:* 539–543 (1981).
10. Herlyn, M.; Steplewski, Z.; Herlyn, D.; Koprowski, H: Colorectal carcinoma-specific antigen: Detection by means of monoclonal antibodies. Proc. natn. Acad. Sci. USA *76:* 1438–1442 (1979).
11. Koprowski, H.; Steplewski, Z.; Mitchell, K.; Herlyn, M.; Herlyn, D.; Fuhrer, P.: Colorectal carcinoma antigens detected by hybridoma antibodies. Somatic Cell Genet. *5:* 957–972 (1979).
12. Magnani, J. L.; Brockhaus, M.; Smith, D. F.; Ginsburg, V.; Blaszczyk, M.; Mitchell, K. F.; Steplewski, Z.; Koprowski, H.: A monosialoganglioside is a monoclonal antibody-defined antigen of colon carcinoma. Science *212:* 55–56 (1981).

13 DelVillano, B. C.; Brennan, S.; Brock, P.; Bucher, C.; Liu, U.; McMhere, M.; Rake, B.; Space, S.; Westrick, B.; Shoemaker, H.; Zurawski, V. R. Jr.: Radioimmunometric assay for a monoclonal antibody-defined tumor marker, CA 19-9. Clin. Chem. *29:* 549–552.
14 Kleist, v. S.; Chavanel, S.; Burtin, P.: Identification of an antigen from normal human tissue that crossreacts with the carcinoembryonic antigen. Proc. natn. Acad. Sci. USA *69:* 2492–2494.
15 Reynolds, J. C.; Mestvedt, B.; Larson, S. M.; Beaumier, P.; Carrasquillo, J. A.; Hellström, I.; Hellström, K. E.: Measurement of plasma p-97 levels in melanoma patients using a new radioassay. Meeting Abstract, J. nucl. Med. *24:* 35 (1983).

Meenhard Herlyn D. V. M., The Wistar Institute of Anatomy and Biology, 36th at Spruce Street, Philadelphia, PA 19104 (USA)

In Vivo Localization of Polyclonal and Monoclonal Anti-CEA Antibodies in Human Colon Carcinomas

J.-P. Mach[1], F. Buchegger[2], M. Forni[3], J. Ritschard[3], C. Haskell[1], S. Carrel[1], A. Donath[3]

[1] Ludwig Institute for Cancer Research, Lausanne Branch, and the
[2] Department of Biochemistry, University of Lausanne, Epalinges, Switzerland
[3] Department of Internal Medicine and Division of Nuclear Medicine, University Hospital, Geneva, Switzerland

Early Experimental Results

Research on tumour localization of radiolabelled antibodies was initiated almost 30 years ago by *Pressman* [1] and *Bale* [2], who showed that labelled antibodies against Wagner osteosarcoma or Walker carcinoma cells were concentrated in vivo by these tumours.

In 1974, we introduced into this field the model of nude mice bearing grafts of human colon carcinoma and the use of affinity purified antibodies against carcinoembryonic antigen (CEA) [3]. We showed that purified ^{131}I-labelled goat anti-CEA antibodies could reach up to a 9 times higher concentration in the tumour than in the liver, while the concentration of control normal IgG in the tumour was never higher than 2.3 times that in the liver. We observed, however, great variations in the degree of specific tumour localization by the same preparation of labelled antibodies, when colon carcinoma grafts derived from different donors were tested. This is probably due to the fact that human tumours keep their initial histologic properties and degree of differentiation after transplantation into nude mice and these two factors appear to affect the ease with which circulating antibodies gain access to the CEA present in tumours. The detection of ^{131}I-labelled antibodies in tumours by external scanning also gave variable results. With colon carcinoma grafts from certain donors we

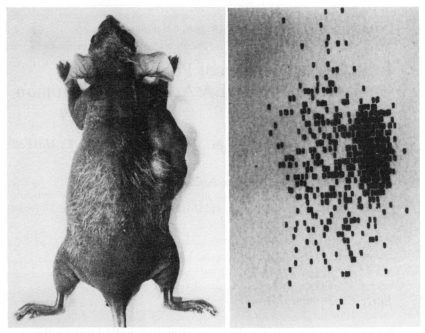

Fig. 1. Scanning of a nude mouse which received an injection of ^{131}I-labelled anti-CEA antibodies. *(A)* Nude mouse bearing a xenograft of human colon carcinoma shown in the scanning position; *(B)* The total body scan from the same mouse obtained 3 days after injection of 2 μg of ^{131}I-labelled anti-CEA antibodies (dose of radioactivity injected = 16 μCi). (Reproduced with the permission of Nature [3]).

obtained scans with good tumour localization, such as the one presented in figure 1, whereas with colon carcinoma grafts from other donors the antibody uptake was not sufficient to give satisfactory scanning images. In this context we think that results in the nude mouse model are a good reflection of the clinical reality observed in patients.

Independently, *Goldenberg et al.* [4] showed specific tumour localization and detection by external scanning with ^{131}I-labelled IgG fractions of anti-CEA serum, using two human carcinomas which had been serially transplanted into hamsters for several years. Using the same experimental model *Hoffer et al.* [5] also demonstrated tumour localization with radiolabelled IgG anti-CEA by external scanning.

Clinical Results with Polyclonal Anti-CEA Antibodies

The first detection of carcinoma in patients obtained by external scanning following injection of purified ^{131}I-labelled anti-CEA antibodies was reported by *Goldberg et al.* [6, 7]. They claimed that almost all the CEA producing tumours could be detected by this method and that there were no false positive results. However, our experience, using highly purified goat anti-CEA antibodies and the same blood pool subtraction technology as *Goldberg* was that only 42% of CEA producing tumours (22 out of 53 tested) could be detected by this method [8, 9]. Furthermore, we found that in several patients the labelled anti-CEA antibodies localized non-specifically in the reticuloendothelium. Despite the use of the subtraction technology, these non-specific uptakes were difficult to differentiate from the specific uptakes corresponding to the tumours. This discrepancy of results is unlikely to be due to a difference in the quality of the anti-CEA used, since we showed by direct measurement of the radioactivity in tumours resected after injection, that our antibody was capable of excellent tumour localization [8] (fig. 2). Furthermore, in a few patients scheduled for tumour resection, we injected simultaneously 1 mg of goat anti-CEA antibodies labelled with 1 mCi of ^{131}I and 1 mg of control normal goat IgG labelled with 0.2 mCi of ^{125}I. By this paired labelled method adapted to the patient situation, we could demonstrate that the antibody uptake was 4 times higher than that of control normal IgG [8] (fig. 2).

These results were very encouraging in terms of specificity of tumour localization. However, the direct measurement of radioactivity in tumours also showed that only 0.05–0.2% of the injected radioactivity (0.5–2 µCi out of 1000 µCi) were recovered in the resected tumours 3–8 days after injection [8]. This information is essential if one is considering the use of radiolabelled antibody for therapy [10].

Monoclonal Anti-CEA Antibody Used in Photoscanning

The obvious advantage of monoclonal antibodies (Mabs) are their homogeneity and their specificity for the immunizing antigen. Another advantage of Mabs is that they each react with a single antigenic determinant and thus should not be able to form large immune

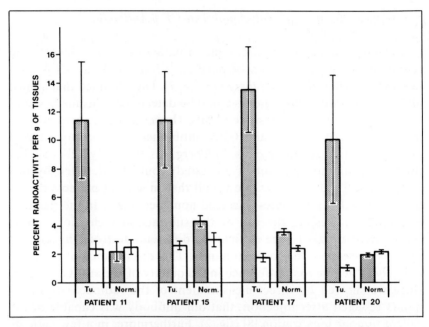

Fig. 2. Specific tumour uptake of anti-CEA antibodies in patients with colon carcinoma. The shaded areas show the relative concentration of ^{131}I-labelled control normal goat IgG, in both tumour (Tu.) and normal mucosa (Norm.) from four patients who received simultaneous injections of both labelled proteins 3–8 days before surgery. The vertical solid lines show the standard deviation of the results obtained in individual tissue fragments. The radioactivity of each isotope present in each fragment was measured in a dual channel gamma counter. (Reproduced with the permission of New Engl. J. Med. [8]).

complexes with the antigen (provided that the antigenic determinant is not repetitive).

The first Mab anti-CEA used for immunoscintigraphy in patients was Mab 23 [11]. Its production and characteristics have been described [12]. Mab 23 was given intravenously to 26 patients with large bowel carcinomas and 2 patients with pancreatic carcinomas. Each patient received 0.3 mg of purified Mab labelled with 1–1.5 mCi of ^{131}I. The patient's premedication included lugol 5% iodine solution, promethazine and prednisolone, as previously described [8, 9]. The patients had no personal history of allergy. They were also tested with an intracutaneous injection of normal mouse IgG and found to have no hypersensitivity against this protein. None of the patients showed any sign of discomfort during or after the injection of labelled mouse antibodies. The patients were studied by external photoscanning 24, 36, 48 and 72 h after injection. An Elscint large-field camera with an LFC9 high-energy parallel-hole collimator was used. In 14 of the 28 patients (50%) a radioactive spot corresponding to the tumour was detected 36–48 h after injection. In 6 patients the scans were doubtful and in the remaining 8 patients they were entirely negative [11].

In order to improve these results we produced a series of 26 new hybrids secreting anti-CEA antibodies. Three of them were selected by criteria of high affinity for CEA of the antibody produced [13]. The new selected Mabs designated 202, 35 and 192 were purified and tested for the detection of human carcinoma grafted in nude mice both in the form of intact Mab and in the form of F(ab')$_2$ and Fab fragments. The results showed that the fragment of Mabs were markedly superior to the intact Mab for the detection of human carcinoma in the nude mouse model [14].

One of the positive scanning studies obtained recently with a new anti-CEA Mab (clone 115) [13] is illustrated in figure 3. The patient was a 77-year old female with a poorly differentiated Dukes C carcinoma of the right colon. The asymptomatic tumour was discovered by barium enema following the observation of an unexplained anemia. The serum CEA level was only 1.5 ng/ml. Figure 3, panel A shows the anterior scan of the abdomen and pelvis, detecting the ^{131}I radioactivity 48 h after injection of 0.5 mg of intact Mab 115 labelled with

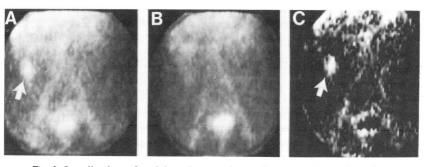

Fig. 3. Localization of a right colon carcinoma by external anterior photoscanning after injection of ^{131}I-labelled monoclonal anti-CEA antibody. *(A)* scan of the abdomen and pelvis detecting ^{131}I radioactivity 48 h after injection of the labelled antibody; *(B)* scan detecting ^{99m}Tc (taken in the same position) 15 min after injection of ^{99m}Tc labelled albumin and free ^{99m}TcO$_4$; *(C)* shows the remaining ^{131}I radioactivity after the subtraction of ^{99m}Tc radioactivity. The radioactive spot corresponding to the tumour is marked by an arrow (*A* and *C*).

1.5 mCi of ^{131}I (55.5 MBq). In addition to the major radioactive uptakes in the upper part and in the middle lower part of the scan, corresponding to the liver and urinary bladder, one sees a distinct asymetric spot in the left part of the scan (arrow) corresponding to the tumour. Panel B shows the scan taken in the same position but detecting ^{99m}Tc 15 min after injection of 500 µCi of ^{99m}Tc labelled human serum albumin and 500 µCi of free ^{99m}TcO$_4$. One sees the radioactivity corresponding to the liver and bladder but no radioactive uptake in the tumour area. This non-specific blood pool and secreted radioactivity can be subtracted from the radioactivity associated with the antibodies. Panel C shows the remaining ^{131}I radioactive uptakes after computerized subtraction of ^{99m}Tc radioactivity. Here the radioactive spot corresponding to the tumour (arrow) is well contrasted.

Four days after the antibody injection, the patient was operated on and an ulcerated carcinoma of 7 cm diameter, weighing about 25 g was resected. The concentration of ^{131}I radioactivity in the tumour was 2.5 times higher than in the dissected adjacent normal mucosa, 5.3 times higher than in the total bowel wall after separation of the mucosa and 17 times higher than in adjacent normal fat.

Tumor Detection by Tomoscintigraphy

Another way to improve tumour detection by immunoscintigraphy is the use of tomoscintigraphy. As we have seen, static photoscanning is limited in part by the presence of radiolabelled antibodies or free ^{131}I released from them, in the circulation, the reticuloendothelial system, the stomach, intestine and urinary bladder. Increased radioactivity in these compartments may give false positive results. Specific tumour sites may be masked by non-specific radioactivity. These problems cannot be entirely resolved by the presently available subtraction methods using ^{99m}Tc labelled HSA and free ^{99m}TcO$_4$. Axial transverse tomoscintigraphy is a method initially developed by *Kuhl and Edwards* in 1973 [15] with the potential to resolve some of these problems. This method, also called single photon *emission* computerized tomography (SPECT), corresponds to the application of the tomographic technique used in *transmission* computerized axial tomography (CAT) to scintigraphic data. Mathematical techniques similar to those used in positron and X-ray tomographies allow the reconstruction of transverse sections as well as frontal, saggital or oblique sections of patients. In collaboration with *C. Berche* and *J.-D. Lumbroso* from the Institut Gustave Roussy in Villejuif, we have recently shown that tomoscintigraphy can improve the sensivity and specificity of tumour detection by radiolabelled anti-CEA Mabs [16]. With this method 15 out of 16 carcinoma tumour sites studied (including 10 colorectal carcinomas, 1 stomach, 1 pancreas and 4 medullary thyroid carcinomas) were detectable. These results are encouraging in terms of sensitivity. However, it should be noted that numerous non-specific radioactive spots, sometimes as intense as the tumours were observed. Thus, the problem of non-specific accumulation of antibodies remains, but the three dimensional localization of radioactive spots by tomoscintigraphy, helps to discriminate specific tumour uptakes from the non-specific ones [16].

Discussion

It is evident from this brief review of our recent results that the method of immunoscintigraphy for the detection of solid tumours can be improved by the use of Mabs selected for higher specificity and affinity for CEA. Other possibilities of improvement of immunoscintigraphy included the use of Mabs against newly discovered tumour markers and the use of other isotopes for the radiolabelling of Mabs. In collaboration with *Mach et al.* [17], we have recently shown that a Mab (17-1A) directed against a new colorectal carcinoma marker can be used for the detection by scintigraphy of these tumours in patients. Other groups have used a Mab raised against human osteosarcoma to detect colorectal carcinomas [18] or Mabs against milk fat globule antigens to detect human carcinoma of various origins [19]. It is clear that all these results need to be confirmed on well controlled clinical studies before they can be recommended for large scale use.

Concerning new isotopes, chelates, such as diethylenetriamine pentaacetic acid (DTPA), have been used to label antibodies with ^{111}Indium [20, 21] which in terms of specific energy and physical half-life is particularly suitable for immunoscintigraphy. An advantage of DTPA is that it can also be used to label antibodies with different alpha-emitting isotopes which represent the best potential agents for the destruction of tumour cells [21]. Critical experimental investigations are, however, necessary, before such types of radioimmunotherapy can be considered for the treatment of cancer patients.

References

1 Pressman, D.; Korngold, L.: The in vivo localization of anti-Wagner osteogenic sarcoma antibodies. Cancer *6:* 619–623 (1953).
2 Bale, W. F.; Spar, I. L.; Goodland, R. L.; Wolfe, D. E.: In vivo and in vitro studies of labeled antibodies against rat kidney and Walker carcinoma. Proc. Soc. exp. Biol. Med. *89:* 564–568 (1955).
3 Mach, J.-P.; Carrel, S.; Merenda, C.; Sordat, B.; Cerottini, J.-C.: In vivo localisation of radiolabelled antibodies to carcinoembryonic antigen in human colon carcinoma grafted into nude mice. Nature *248:* 704–706 (1974).
4 Goldenberg, D. M.; Preston, D. F.; Primus, F. J.; Hansen, H. J.: Photoscan localization of GW-39 tumors in hamsters using radiolabelled anti-carcinoembryonic antigen immunoglobulin G. Cancer Res. *34:* 1–9 (1974).
5 Hoffer, P. B.; Lathrop, K.; Bekerman, G.; Fang. V. S.; Refetoff, S.: Use of ^{131}I-CEA antibody as a tumor scanning agent. J. nucl. Med. *15:* 323–327 (1974).
6 Goldenberg, D. M.; Deland, F.; Enishin, K.; Bennett, S.; Primus, F. J.; Nagell, J. R. van; Estes, N.; DeSimone, P.; Rayburn, P.: Use of radiolabelled antibodies to carcinoembryonic antigen for the detection and localization of diverse cancers by external photoscanning. N. Engl. J. Med. *298:* 1384–1388 (1978).
7 Goldenberg, D. M.; Kim, D. D.; DeLand, F. H.; Bennett, S.; Primus, F. J.: Radioimmunodetection of cancer with radioactive antibodies to carcinoembryonic antigen. Cancer Res. *40:* 2984–2992 (1980).
8 Mach, J.-P.; Carrel, S.; Forni, M.; Ritschard, J.; Donath, A.; Alberto, P.: Tumour localization of radiolabelled antibodies against carcinomebryonic antigen in patients with carcinoma. N. Engl. J. Med. *303:* 5–10 (1980).
9 Mach, J.-P.; Forni, M.; Ritschard, J.; Buchegger, F.; Carrel, S.; Widgren, S.; Donath, A.; Alberto, P.: Use and limitations of radiolabelled anti-CEA antibodies and their fragments for photoscanning detection of human colorectal carcinomas. Oncodevelop. biol. Med. *1:* 49–69 (1980).
10 Order, S. E.; Klein, J. L.; Ettinger, D.; Alderson, P.; Siegleman, S.; Leichner, P.: Use of isotopic immunoglobulin in therapy. Cancer Res. *40:* 3002–3007 (1980).
11 Mach, J.-P.; Buchegger, F.; Forni, M.; Ritschard, J.; Berche, C.; Lumbroso, J. D.; Schreyer, M.; Girardet, C.; Accolla, R. S.; Carrel, S.: Use of radiolabelled monoclonal anti-CEA antibodies for the detection of human carcinomas by external photoscanning and tomoscintigraphy. Immunology Today *2:* 239–249 (1981).
12 Accolla, R. S.; Carrel, S.; Mach, J.-P.: Monoclonal antibodies specific for carcinoembryonic antigen and produced by two hybrid cell lines. Proc. natn. Acad. Sci (USA) *77:* 563–566 (1980).
13 Haskell, C. M.; Buchegger, F.; Schreyer, M.; Carrel, S.; Mach J.-P.: Monoclonal antibodies to carcinoembryonic antigen: ionic strength as a factor in the selection of antibodies for immunoscintigraphy. Cancer Res. *43:* 3857–3864 (1983).
14 Buchegger, F.; Haskell, C. M.; Schreyer, M.; Scazziga, B. R.; Randin, S.; Carrel, S.; Mach, J.-P.: Radiolabelled fragments of monoclonal anti-CEA antibodies for

localization of human colon carcinoma grafted into nude mice. J. exp. Med. *158:* 413–427 (1983).
15 Kuhl, D. E.; Edwards, R. D.: Image separation radioisotope scanning. Radiology *80:* 653–662 (1963).
16 Berche, C.; Mach, J.-P.; Lumbroso, J.-D.; Langlais, C.; Aubry, F.; Buchegger, F.; Carrel, S.; Rougier, P.; Parmentier, C.; Tubiana, M.: Tomoscintigraphy for detecting gastrointestinal and medullary thyroid cancers: first clinical results using radiolabelled monoclonal antibodies against carcinoembryonic antigen. Br. med. J. *285:* 1447–1451 (1982).
17 Mach, J.-P.; Chatal, J.-F.; Lumbroso, J.-D.; Buchegger, F.; Forni, M.; Ritschard, J.; Berche, C.; Douillard, J.-Y.; Carrel, S.; Herlyn M.; Steplewski, Z.; Koprowski, H.: Tumour localization in patients by radiolabelled antibodies against colon carcinomas. Cancer Res (in press).
18 Farrands, P. A.; Pimm, M. V.; Embleton, M. J.; Perkins, A. C.; Hardy, J. D.; Baldwin, R. W.; Hardcastel, J. D.: Radioimmunodetection of human colorectal cancers by an anti-tumour monoclonal antibody. Lancet *ii:* 397–400 (1982).
19 Epenetos, A.-A.; Mather, S.; Granowska, M.; Nimmon, C. C.; Hawkins, L. R.; Britton, K. E.; Shepherd, J.; Taylor-Papadimitriou, J.; Durbin, H.; Malpas, J. S.; Bodmer, W. F.: Targeting of iodine-123-labelled tumour associated monoclonal antibodies to ovarian, breast and gastrointestinal tumours. Lancet *ii:* 999–1005 (1982).
20 Khaw, B.; Fallon, J. T.; Strauss, H. W.; Haber, E.: Myocardial infarct imaging of antibodies to canine cardiac myosin with Indium-111-diethylenetriamine pentaacetic acid. Science *209:* 295–297 (1980).
21 Scheinberg, D. A.; Strand, M.; Gansow, O. A.: Tumour imaging with radioactive metal chelates conjugated to monoclonal antibodies. Science *215:* 1511–1513 (1982).

Dr. J.-P. Mach, Ludwig Institute for Cancer Research, CH-1066 Epalinges S/Lausanne (Switzerland)

Immunotherapy of Gastrointestinal Malignancies with a Murine IgG 2A Antibody*

H. F. Sears[1], D. Herlyn[2], H. Koprowski[2], J. W. Shen[1]

[1] The Fox Chase Cancer Center, Philadelphia, Pa., USA
[2] The Wistar Institute of Anatomy and Biology, Philadelphia, Pa., USA

Introduction

A murine monoclonal antibody (MCA) was developed by the Wistar Institute of Anatomy and Biology (Philadelphia, PA, USA) from a hybridoma of splenocytes immunized by a culture line of human colon adenocarcinoma (SW1083). The resultant MCA was of the gamma 2A isotype [1] and bound to 8 of 9 human colorectal cancer (CRC) tissue culture lines and not to a variety of other human tumour cell cultures, fibroblasts, peripheral blood leukocytes, or red blood cells [2]. It bound to freshly resected CRC specimens and not to the surrounding normal colonic mucosa. This antibody, 1083-17-1A, mediated specific lysis of human CRC cells and not other cell types by cell dependent, rather than complement mediators [3]. Mouse lymphocytes and peritoneal macrophages and human lymphocytes and circulating peripheral monocytes are effectors of this cytolysis.

The 17-1A MCA specifically inhibited the growth of human CRC in athymic mice and not other human tumours. This function was not mediated by a complement or NK cells and was abrogated when macrophage function was suppressed by silica administration [4]. The MCA caused complete tumour suppression if given to the mouse 1 or 2 days after injection of 5 million live tumour cells, but not if the interval between tumour cell innoculation and antibody administration were extended beyound this.

* A manuscript describing these patients has been submitted to J. Am. med. Ass.

This antibody was tested in a ex vivo perfusion model using freshly resected human colons containing a cancer designed to assess the binding characteristics and specificity of murine gamma 2A immunoglobulin (Ig) after intravascular administration. The 17-1A and a variety of control MCA's that bound to other non-related antigens were tested with perfusion times of 1–2 h. Intravenously administered MCA's of the gamma 1 and 2A isotypes would exit the circulation and bind specifically to antigens contained on human colonic cancers and the normal mucosa, while IgM's would not, in this model [5]. MCAs to non-relevant antigens did not bind. The 17-1A MCA bound specifically to the colon cancer in ⅓ of the trials, but in ⅓ of the trials it bound to normal human colonic mucosa, as well as to the tumour [6].

The antigen recognized by the 17-1A MCA has not been completely characterized. It is destroyed when the cell membrane is disrupted either by ionic (KCL) or physical means. It is altered by protease, but not neurominadase. It may be a structural membrane protein.

The antigen is not shed into the supernatant of cell cultures of malignant cells that express the antigen on their membranes. It is not found in the circulation of any patient with colon or rectal cancers. It does not cross react with other known GI carcinoma antigens (i.e., CEA or CA 19-9).

Because of the labile nature of the antigen, (destruction by formalin fixation) early attempts to identify the antigen on human tissues using immunoperoxidase techniques were not successful. Recently, *Shen et al.* [7] modified these techniques to document the presence of the MCA 17-1A antigen in a variety of normal human GI mucosal epithelia as well as the malignancies from these sites, both primary and metastatic deposits (fig. 3).

A clinical trial of immunotherapy using the MCA (17-1A) for patients with colonic carcinoma was devised to ask questions about tumour effect, binding specificity, acute (anaphylaxis) and chronic (serum sickness) toxicity, duration of detectable MCA in the human circulation, persistance of activity of antibody function while in the human, and the development of human anti-mouse immunoglobulin response to a systemically administered murine MCA [8, 11]. The requirements of any humorally mediated immunotherapy are
(1) the use of anti-sera of high affinity, specificity and concentration,

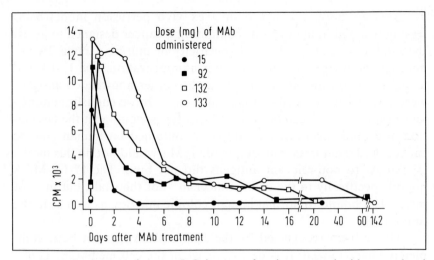

Fig. 1. Persistence of mouse IgG in sera of patients treated with monoclonal antibody. Patients' serum diluted 1:10 in buffer was exposed to rabbit anti-mouse IgG antibody and the binding was detected by ^{125}I labelled rabbit anti-mouse F(ab')$_2$ immunoglobulin. Present data obtained with four individuals who received low or medium doses of MCA 1083-17-1A.

(2) clearance of circulating antigen prior to administering the antibody intended to localize on the tumour tissue and
(3) application of antibody to a state of minimal disease to ensure maintenance of antibody excess for significant periods.

The MCA 17-1A offered an ideal candidate for such trial as it met many of these characteristics, but had to be tested in advanced disease patients prior to use in the ideal candidate, those with minimal disease. Specificity of the 17-1A antibody for human colon cancers and not other tissues was further demonstrated by *Chatal et al.* [9] and *Moldofsky et al.* [10] who documented by isotope scanning selective binding of F(ab')$_2$ fragments of MCA 17-1A to the hepatic and other metastatic deposits of colon carcinoma in patients.

Patients

Our initial immunotherapy trial selected ten patients with advanced gastrointestinal adenocarcinoma whose disease had prompted the need for an intra-abdominal operation because of some complication, most often intestinal obstruction. A variety of other treatment modalities were or had been employed on this group of patients. After skin tests for hypersensitivity and recall antigen were made, patients were given the MCA intravenously or in special cases, intra-arterially 12–48 h prior to the planned laparotomy. In this manner we hoped to obtain tumour specimens to document binding of systemically administered antibody and other tissues to assess the specificity of this binding. We also studied the persistance of mouse Ig (Immunoglobulin) in the patient's serum, the continued functional activity

Table I. Immunotherapy with 17-1A MAb

Name	Sex	DOB	Primary site	Date MAb received	Amount MAb received Mg	Current status	Human anti-mouse antibody
PB	M	9-25	Rectal	12-04-80	15	AWD	+
EH	F	11-05	Colon	2-18-81	150	D	–
JC	M	11-22	C	7-27-81	150	C (Resp)	+
PK	M	9-26	Gastric	9-01-81	200[a, b]	D[d]	+
ME	F	10-11	C	11-03-81	100[a]	D	+
NV	M	5-41	G	12-14-81	72	D	N. T.
NH	F	2-24	R	12-15-81	128	NED	+
EP	M	5-24	R	12-22-81	92[a, b]	AWD	+
JS	M	3-07	C	3-18-82	133[b]	NED	+
AC	M	12-12	C	5-04-82	135[b]	D	?
RH	M	3-17	C	5-11-82	137[a, b]	D	+
WS	M	5-30	C	6-15-82	366	AWD	–
RC	F	6-16	C	8-25-82	440[b]	NED	–
NA	M	10-29	Pancreas	9-22-82	433[b]	NED	–
FS	M	8-17	C[c]	10-25-82	400[b]	D	–
JB	M	9-18	R	10-25-82	380[b]	AWD	–
NB	M	7-06	C	11-01-82	675[b]	D	+
MB	F	7-02	C	12-01-81	391[b]	AWD	–
MJ	M	10-15	C	1-10-83	1000[b]	AWD	–
EK	M	8-30	C	1-10-83	611[b]	AWD	–

AWD = Alive with disease; DOB = Date of birth; NED = No evidence of disease; D = dead.
[a] Intrahepatic artery infusion
[b] MAb mixed with autologous leukocyte-rich plasma
[c] Two primary colon carcinomas
[d] Received multiple doses of MAb

of Ig in the circulation and formation by the host of anti-mouse immunoglobulin antibody over the next year. The requirements for an operative procedure was dropped after ten patients had been treated, and after which another ten patients were treated using a similar schema (table I).

Results

Escalating doses from 15 mg to 1 g of column purified pyrogen free murine IgG were systemically administered over 1 h to patients without any acute pulmonary or cardiac (anaphylactic) reactions. Two patients developed scattered urticarial lesions towards the end of the infusion which resolved spontaneously. If infusion rates increased significantly above that recommended, transient mild hypotension was seen in two patients getting the large mg amounts of MCA which corrected as the infusion rate was slowed. No evidence of chronic toxicity was noted. One patient developed a rash 12 days after antibody exposure which resolved spontaneously, and one patient, the first treated with 15 mg, had 1 day of neuritis like symptoms 3 weeks after antibody administration which had resolved spontaneoulsy by the next day when he was examined. Fourteen patients had antibody mixed with plasmaphoresed autologous leukocyte suspension which did not alter the pattern of toxicity or immunization.

One patient who received 5 injections over 11 days, developed an anaphylactic reaction during his 5th exposure to MCA. Subsequently, only one injection of murine Ig was given to each patient, until recently when repeated exposure has been tested again in certain specifically prepared patients.

Murine MCA could be detected in patients' circulation from 2–26 days after administration depending upon the dose given and the subsequent development of human anti-mouse immunoglobulin antibody (figs. 1 and 2). Actual function or specific binding to CRC target cells of the circulating antibody decayed much more quickly and seemed to be gone within 7 days despite the initial mg amount administered. The effect of tumour burden, which might bind significant amounts of MCA and remove it from the systemic circulation, could not be quantified as there is no objective measure of tumour burden in these patients. The route of administration or pre-mixing of

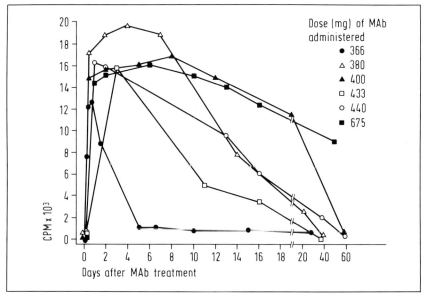

Fig. 2. Presents data obtained with six individuals who received large doses of MCA 1083-17-1A.

antibody with autologous leukocytes did not affect the decay of antibody titers or the development of allergy.

Patients who received 200 mg or less of MCA, who were not anergic because of their disease or its treatment, developed human anti-mouse immunoglobulin antibody. This could usually be detected by the 10–14th day after administration and would decay over the next 2–3 months. It has persisted for much longer in isolated situations. When a human anti-mouse antibody is found, anaphylaxis is possible in as little as 11 days from the initiation of therapy.

Large initial amounts of murine Ig, greater than 500 mg, seemed to induce tolerance to MCA, as in all but two of the patients so treated, no measurable amounts of human anti-mouse immunoglobulin antibody have been detected (table II). Those two patients that did develop an anti-mouse Ig antibody response had previously been skin tested and scanned using a F(ab')$_2$ fragment of this MCA for an imaging trial.

Binding specificity assays using the immunoperoxidase technique have recently documented that systemically administered MCA can

Table II. Correlation between dosage of MAb and human anti-mouse antibody response

Amount of mouse MAb given (mg) in single injection	Ratio of patients developing human anti-mouse antibody
15–200	8[a]/9
366–1,000	1[b]/9

[a] One immunosuppressed patient who received 150 mg of MAb did not develop human anti-mouse antibodies
[b] Patient who received 675 mg showed presence of human anti-mouse antibodies only between the 14th and 37th day after treatment

be detected on tumour metastases 24–48 h after intravenous infusion (fig. 3). At 7–8 days bound antibody is either marginally detectable or extremely focal in its distribution on these tumours. By three weeks it is completely gone, though the antigen is still expressed on the cell surface and will bind 17-1A antibody that tissue is incubated with at that time. We have no evidence of antigen modulation. Systemically administered MCA also was detected bound to the normal colonic mucosal epithelium at 24–48 h (fig. 4), but not thereafter, though again the antigen persist.

Tumour effect is extremely difficult to assess in this group of variably treated patients who had multiple presentations of complications of their disease. Eleven of the 20 patients are alive 6–31 months after treatment. Of those, four currently have no evidence of their disease being present 10–18 months since the single exposure to MCA 17-1A, while in the remaining seven, the metastatic disease is progressing. Three of the patients who are not currently tumour free or who have died, have had alterations in their tumour that were temporally associated with the MCA exposure (table III). The first patient treated, given 15 mg, had no detectable tumour for 18 months after treatment, while another patient had marked disease reduction when MCA was added to a chemotherapeutic regimen. Two of the patients, who are currently free of any evidence of tumour were treated with chemotherapy or radiation therapy that they had not previously had, in conjunction with the MCA immunotherapy, while the other two patients had already failed chemotherapy and had no other treatment since the MCA was administered. These patients did not

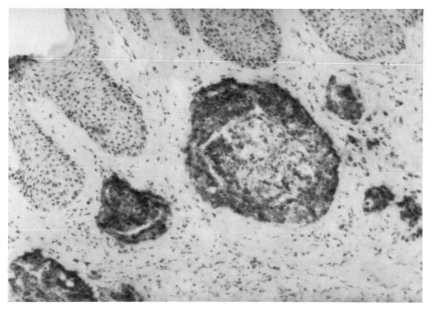

Fig. 3. Photomicrograph x 100 of a subcutaneous metastasis of a colon carcinoma showing bound mouse immunoglobulin 48 h after intravenous administration of 700 mg. IgG 17-1A.

have huge tumour burdens at the time of inital therapy and probably represent the ideal candidate for this type of treatment. One of these patients treated 18 months ago had a pulmonary metastases and a pelvic lymph node metastases resected while a residual pelvic mass remained as an indicator lesion. She currently has a normal chest X-ray and CAT scan of her abdomen and pelvis and is symptomatically symptom free. Her tumour markers were never markedly elevated, but have been absolutely normal since she was treated.

Two patients, both with progressive tumours (one at 1 month the other at 12 months) who initially received large amounts of MCA and did not develop detectable human anti-mouse Ig antibody have subsequently received a second intravenous injection of 17-1A. Neither had any acute or chronic reaction. The second exposure was a smaller (200 mg) or immunogenic dose. It broke tolerance in one patient, who 14 days after the second injection had circulating human anti-mouse Ig. The material from the second patient has not been tested.

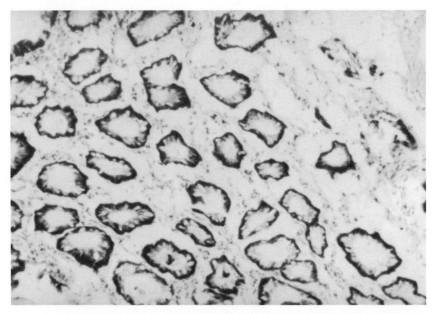

Fig. 4. Photomicrograph x 100 of normal colon mucosa 48 h after intravenous adminstration of 100 mg of IgG 17-1A.

Discussion

We have studied the effect of immunotherapy with mouse monoclonal antibody in a human solid tumour system using the 1083-17-1A MCA which binds to human colorectal cancers (CRC) maintained in tissue culture [1] and is tumouricidal for all CRC which have been implanted in nude mice [3]. The results of imaging of CRC in humans with this radiolabelled monoclonal antibody seem to indicate that the label can be detected in the tumour and in no other tissue on gamma scintigraphy [9, 10]. Up to 1 g of this purified monoclonal antibody can be given safely to patients by intravenous infusion without any side effects, and the mouse IgG is detectable in patients' circulation from 2–50 days, depending on the dose administered. When the dose of monoclonal antibody was 200 mg or less, all patients developed anti-mouse antibodies which were present in the circulation for considerable lengths of time. When the dose of mono-

Table III. Evaluation of immunotherapy with MAb

Patients	Site of primary/ differentiation	Site of metastases	Therapy prior to MAb	Operation	MAb given[a] in mg	Therapy post MAb	Mos. NED	CEA (ny/ml)	Recurrence site post MAb	Current status (Mos.)
PB	R/M	Pelvic Colonic Obstruction	None	Colostomy	15	None	18	N.T.	Local Bone Lung	Alive Lung mets. (32 Mos.)
EP	R/M	Liver	Chemo. & Radiation	Liver Resection	92	None	8	17.3 PreAb 1.6 3d 2.6 60d 2.9 90d 7.7 180d	Liver	Alive Liver mets. (18 Mos.)
NH	R/M	Lung Pelvic	Chemo.	Thoracotomy Ischiorectal Biopsy	125	None	12+	5 PreAb <2.5 S.nce	None	A&W NED- (18 Mos.)
JS	C/M	Local Retro- peritoneal	None	Colostomy	133	Radiation	9+	25 PreAb < 2.5	None	A&W NED- (15 Mos.)
JC	C/P	Liver Retro- peritoneal Ureteral Colonic Obstruction	Chemo.	Ileo- transverse Colostomy	150	Chemo.	0	23.6 PreAb 48.0 30d 6.7 60d 2.0 90d 75 180d	Liver Retro- peritoneal	Died (12 Mos.)
WS	C/P	Peritoneal	Chemo.	None	366	None	9	4.5 PreAb <2.5 30d	None	Alive Peritoneal mets. (12 Mos.)
RC	C/M	Lung Liver	None	None	440	Chemo.	3+	144.8 PreAb 30.8 60d 12.5 90d 10.0 ≥50d	None	A&W-NED (10 Mos.)
NA	P_r/M	Liver	Chemo.	None	433	Chemo.	3+	Never elevated	None	A&W-NED (9 Mos.)

[a] Within 3–48 h before the operation.

R = Rectum; M = Moderately well differentiated; A&W = Alive and well; NED = No evidence or disease;
C = Colon; P = Poorly differentiated; N.T. = Not tested; CEA = Carcinoembryonic antigen;
P_r = Pancreas; Mos. = Months

clonal antibody given exceeded 366 mg, patients who were otherwise capable of mounting an immune response did not develop antimouse antibodies. Whether we have induced tolerance to a foreign immunogenic protein in a patient who is not otherwise immunosupressed remains to be clarified by further study.

Eleven patients treated in the initial phases of the study had widespread metastases which compromised hepatic or renal function, and nine of those also had an abdominal operative procedure. In this clinical setting, we were limited to observations indicating the absence of acute toxicity of intravenously administered monoclonal antibody and the absence of aggravation of these abnormalities after antibody treatment.

Seven patients whose metastases had not caused hepatic or renal dysfunction are alive 10–32 months after treatment. Four of these seven patients who received 125–440 mg of monoclonal antibody have no evidence of residual disease (NH, JS, RC and NA, table II). In five of these patients, CEA levels were elevated prior to administration of monoclonal antibody, and subsequently decreased in all five, reaching normal values. Three of these patients failed prior chemotherapy. Although radiation therapy in one patient (JS) and chemotherapy in another (RC) after monoclonal antibody treatment preclude an exact assessment of the role of immunotherapy in these outcomes, one patient (NH, table I) who failed previous chemotherapy and received no other treatment following immunotherapy has no evidence of disease 18 months after monoclonal antibody treatment.

Immunotherapy with a monoclonal antibody that has been proven to specifically destroy human colorectal cancer xenotransplants in mice, may inhibit gastrointestinal cancer in patients with metastatic disease. These results warrant use of this monoclonal antibody in more extensive clinical trials with patients who have limited tumour burdens. Treatment with monoclonal antibody could also be considered in conjunction with chemotherapy of tumours of the gastrointestinal tract.

Summary

Twenty patients with advanced gastrointestinal adenocarcinoma were given a murine monoclonal antibody that specifically binds to and kills human colon adenocarcinoma *in vitro* and *in vivo*. Dosages of 15–1,000 mgs of antibody were well tolerated and could be detected in the patient's circulation for cariable periods up to 60 days after administration. Patients receiving Ig antibody which was dected between 10 and 14 days after injection and could be found for as long as 360 days in one patient. Patients receiving more than 300 mgs, as their first exposure to murine Ig, did not make measurable anti-mouse antibody. Some patients had singificant tumour reduction in association with monoclonal antibody immunotherapy, but associated clinical therapies make causal relationships difficult to assess.

References

1 Koprowski, H.; Steplewski, Z.; Mitchell, K.; Herlyn, M.; Herlyn, D.; Fuhrer, P.: Colorectal carcinoma antigens detected by hybridoma antibodies. Somatic Cell Genet. *3:*957–972 (1979).
2 Herlyn, M. F.; Steplewski, Z.; Herlyn, D. M.; Koprowski, H.: Colorectal carcinomaspecific antigen: Detection by means of monoclonal antibodies. Proc. natn. Acad. Sci. USA *76:*1438–1442 (1979).
3 Herlyn, D. M.; Steplewski, Z.; Herlyn, M. F.; Koprowski, H.: Inhibition of growth of colorectal carcinoma in nude mice by monoclonal antibody. Cancer Res. *40:*717–721 (1980).
4 Herlyn, D. M.; Koprowski, H.: IgG2a monoclonal antibodies inhibit human tumor growth through interaction with effector cells. Proc. natn. Acad. Sci. USA *79:* 4761–4765 (1982).
5 Sears, H. F.; Steplewski, Z.; Herlyn, M.; Herlyn, D.; Grotzinger, P.; Koprowski, H.: Ex vivo perfusion of human colon with monoclonal anticolorectal cancer antibodies. Cancer *49:*1231–1236 (1982).
6 Sears, H. F.; Herlyn, D.; Herlyn, M.; Grotzinger, P.; Steplewski, Z.; Gerhard, W.; Koprowski, H.: Ex vivo perfusion of a tumor-containing colon with monoclonal antibody. J. surg. Res. *31:*145–150 (1981).
7 Shen, J. W.; Atkinson, B.; Koprowski, H.; Sears, H. F.: Binding of murine immunoglobulin to human tissues after immunotherapy with anticolorectal carcinoma monoclonal antibody. Int. J. Cancer (Submitted).
8 Sears, H. F.; Atkinson, B. F.; Matthis, J.; Herlyn, D.; Ernst, C.; Mattis, J.; Steplewski, Z.; Hayry, P.; Koprowski, H.: The use of monoclonal antibody in a Phase I clinical trial of human gastrointestinal tumors. Lancet *1:*762–765, (1982).
9 Chatal, J. F.; Saccavini, J. C.; Fumoleau, P.; Bardy, J.; Douillard, J.; Aubry, B.; LeMevel, U.: Photoscanning localization of human tumours using radioiodinated monoclonal antibodies to colorectal carcinoma. Proceedings of the 29th Annual Meeting of the Society of Nuclear Medicine (Abstract). J. nucl. Med. *23:* P8 (1982).
10 Moldofsky, P. J.; Powe, J.; Mulhern, C. B.; Sears, H. F.; Hammond, N.; Gatenby, R.; Steplewski, Z.; Koprowski, H.: Imaging with radiolabelled F(ab')$_2$

fragments of monoclonal antibody in patients with gastrointestinal carcinoma. Radiology *149:* 549–555, (1983).
11 Koprowski, H.: Monoclonal antibodies in vivo; in Langman, Trowbridge, Dulbecco (eds.), Monoclonal antibodies and cancer. Proc. of the 4th Armand Hammer Cancer Symposium. (Academic Press, New York, in press, 1983).

H. F. Sears, M. D., The Fox Chase Cancer Center, Central & Shelmire Aves., Philadelphia, PA 19111 (USA)

THE LIBRARY
UNIVERSITY OF CALIFORNIA
San Francisco
476-2334

THIS BOOK IS DUE ON THE LAST DATE STAMPED BELOW
Books not returned on time are subject to fines according to the Library Lending Code. A renewal may be made on certain materials. For details consult Lending Code.

14 DAY
FEB 20 1986
RETURNED
FEB 18 1986